WŁADYSŁAW KRAJEWSKI

CORRESPONDENCE PRINCIPLE AND GROWTH OF SCIENCE

SPRINGER-SCIENCE+BUSINESS MEDIA, B.V

Library of Congress Cataloging in Publication Data

Krajewski, Wladyslaw.
　　Correspondence principle and growth of science.

　　(Episteme ; v. 4)
　　Bibliography: p.
　　Includes index.
　　　1.　Correspondence principle (Quantum mechanics)
2.　Science — Philosophy.　I.　Title.
QC174.17.C5K7　　　500.2'09　　　76-49497
ISBN 978-94-010-1180-8　　　ISBN 978-94-010-1178-5 (eBook)
DOI 10.1007/978-94-010-1178-5

TABLE OF CONTENTS

PREFACE

This book is devoted to the problems of the growth of science. These problems, neglected for a long time by the philosophers of science, have become in the 60's and 70's a subject of vivid discussion. There are philosophers who stress only the dependence of science upon various sociological, psychological and other factors and deny any internal laws of the development of knowledge, like approaching the truth. The majority rejects such nihilism and searches for the laws of the growth of science. However, they often overlook the role of the Correspondence Principle which connects the successive scientific theories. On the other hand, some authors, while stressing the role of this principle, overlook logical difficulties connected with it, e.g. the problem of the incompatibility of successive theories, of the falsity of some of their assumptions, etc.

I believe the Correspondence Principle to be a basic principle of the progress of contemporary physics and, probably, of every advanced science. However, this principle must be properly interpreted and the above-mentioned logical difficulties must be solved. Their solution requires, as it seems, revealing the idealizational nature of the basic laws of science, in any case of the quantitative laws of advanced sciences. This point has been recently emphasized by some Polish philosophers, especially in Poznań. Therefore, I devote some space to an analysis of the processes of idealization and factualization in science and to the presentation of the main points of a vivid discussion about these problems which took place during the last years in Poland.

In this light, I examine the concept of reduction stressing mainly the reduction of an idealizational law to a factual one. Then, I discuss different interpretations of the correspondence relation and suggest my own interpretation which may be called a 'renewed implicational' version.

Later on, I present different types of methodological empiricism with emphasis on those which point out the idealizational nature of laws and the role of competition among different theories.

In the later chapters I consider the development of science from a more general point of view. The problem of revolutions and continuity (cumulation) in science is examined. A view is presented that there are two kinds of revolutions in science, neither of which leads to a complete breach of continuity. The problem of truth is analysed. Absolute (exact) truth is rare in science. The quantitative laws and *a fortiori* the theories are relatively (approximately) true in the classical sense (i.e. with respect to reality); they may be absolutely true only in ideal models. However, science gradually progresses through relative truths towards the absolute one.

In the last chapter different approaches to the study of science are discussed: internalistic and externalistic, crudely empirical, aprioristic and idealizational-empirical ones. I stand by the view that the main laws of the development of science are internal; however, a scheme based on these laws is an idealization because different external factors always influence real science.

It is time to explain my philosophical attitude. I associate basic ideas of dialectical materialism (liberated from dogmatism and Hegelian phraseology) with logico-methodological achievements of the analytical philosophy of our century (liberated from positivistic narrowness). And first of all, the methods used in the sciences and the history of knowledge must be seriously examined. In the past decades a movement towards this has occurred from two sides. Many Marxist philosophers now pay more attention to logical analysis, many Western analytical philosophers turn to the study of the methods used in advanced sciences and of the growth of knowledge. Both processes are gradually increasing. Therefore, my prognosis about the future of the philosophy of science (I do not speak about philosophy in general) is optimistic.

Western readers usually do not know much about philosophy in Eastern Europe. For this reason I present in this book some of the achievements of the philosophy of science in Poland and, to an extent, in the U.S.S.R. and other countries. Of course, I do not present all the interesting results but only those that are, according to me, important for the solution of the problem of the growth of knowledge. Naturally, only I am responsible for the picture of this growth presented in this book.

Warsaw, September 1975   W.K.

ACKNOWLEDGEMENTS

I have presented many ideas contained in this book at my seminar on the philosophy of science at Warsaw University and at meetings of the divisions of the Polish Philosophical Society in Warsaw, Poznań and Wrocław. Everywhere discussion followed that helped me in my further work on these subjects. I thank all the participants in these discussions.

I am especially indebted to my colleagues and friends Stefan Amsterdamski, Leszek Nowak, Elżbieta Pietruska, Marian Przełęcki, Jan Such, Irena Szumilewicz, Jan Żytkow and to my son Stanisław Krajewski. They have read—entirely or partially—earlier versions of this book and have made numerous critical remarks which essentially contributed to the improvement of my text. To be sure, I could not take all remarks into account; some of them led in opposite directions.

I also thank Sławomir Magala for correcting my English.

W.K.

SYMBOLS AND ABBREVIATIONS
USED IN THIS BOOK

FORMAL LOGIC

$\neg$	Negation
$\wedge$	Conjunction
$\Rightarrow$	Implication
$\underset{x}{\vee}$	Existential quantifier
$\underset{x}{\wedge}$	General quantifier

SET THEORY

$\varnothing$	Empty set
$\cup$	Union of sets
$\cap$	Intersection of sets
$\rightarrow$	One-sided correspondence between sets
$\leftrightarrow$	One-to-one correspondence between sets
$=$	Identity of sets

PHYSICAL THEORIES

GC	(Ptolomy's) geocentric system
HC	(Copernicus') heliocentric system
KL	Kepler's laws
CM	(Newton's) classical mechanics
TD	Thermodynamics
CSM	Classical statistic mechanics
STR	(Einstein's) special theory of relativity
GTR	(Einstein's) general theory of relativity
RQM	(Dirac's) relativistic quantum mechanics

METHODOLOGY OF SCIENCE

(The symbols introduced in this book. In parentheses—the number of the section in which the symbol is explained.

CP	Correspondence principle	(1.1, 1.3)

CR	Correspondence relation	(1.1, 4.1)
L	Law of science	(2.1)
$F(x) = 0$	Functional dependence between some parameters of x	(2.1)
$C(x)$	Set of conditions in which $F(x) = 0$ takes place	(2.1)
D	Domain of a law (or theory)	(2.1, 5.5)
L_F	Factual law	(2.2)
L_I	Idealizational law	(2.2)
C_F	Factual conditions	(2.2)
C_I	Idealizing conditions (assumptions)	(2.5)
S	Description of the structure of a system	(3.2)
B	Bridge laws	(3.2)
L_1	Earlier (corresponded) law	(4.1)
L_2	Later (corresponding) law	(4.1)
L_1'	L_1 reinterpreted in the light of L_2	(4.1)
L_2^*	'Abstracticized' L_2	(4.6)
aL	Approximation of L	(4.3)
Expl(L)	Set of phenomena explained by L	(4.4)
$V(T)$	Vocabulary of the theory T	(5.5)
$V_2(T_1)$	T_1 Translated into the language of T_2	(5.5)
DI	Degree of inadequateness	(6.5, 8.3)
F	Factual sentence	(8.3)
E	Relative error	(8.3)
TrC	Truth content	(8.3)
MTr	Model truth	(8.4)
MTrC	Model truth content	(8.6)

CHAPTER 1

CORRESPONDENCE PRINCIPLE

1. BOHR'S PRINCIPLE

The Correspondence Principle (CP) appeared for the first time in the old quantum theory of the atom created by Niels Bohr. According to this principle, the quantum theory of the atom and of its radiation passes asymptotically into the classical theory when the quantum numbers increase or, in other words, when we may neglect Planck's constant h.[1]

Bohr is commonly acknowledged to be the author of the CP. However, we meet in the literature different opinions about the date of the birth of this principle. Indeed, it is not easy to state it. The basic ideas of the CP are contained already in three famous papers by Bohr from 1913, in which he formulated his quantum postulates concerning electron orbits in the atom. Bohr did not yet use the word 'correspondence' in these papers, he used the word 'analogy' for the relation between classical and quantum theories. Later he wrote himself that we can find there the first germs of the CP (Bohr, 1922, foreword). The word 'correspondence' was used by Bohr in 1920 and later, especially in 1922, in the Nobel lecture. He spoke about the correspondence between classical and quantum light emission's frequencies and intensities and, more generally, between classical and quantum theories of the atom. In later papers Bohr often mentions the CP. It is, however, remarkable that he has never formulated this principle explicitly (cf. Meyer-Abich, 1965; Lewenstam, 1974; Niedźwiedzki, 1974).

Bohr's principle connects the new theory of atom radiation to the old one. The classical theory appeared — at least in the purely mathematical respect — as a limit case of the quantum theory. Such an approach has a great advantage: the new theory has its empirical support not only in the results of new, specific quantal experiments but also in the results of the whole experience confirming the old theory. The physicists appreciated this principle very highly, sometimes enthusiastically. Sommerfeld called it a 'magic stick' (*Zuberstaub*), Kramers a 'torch' (*Leuchtpunkt*), Guth a 'wonderfully effective' (*wunderbar wirksam*) instrument (cf. Laitko, 1969).

Bohr's principle was used by Heisenberg, Born and others in the creation of quantum mechanics (QM) in 1925–27. The basic equations of QM pass asymptotically into the equations of classical mechanics (CM) when we assume that h tends to zero. E.g., the Schrödinger wave equation passes then into the classical Hamilton-Jacobi equation. A simpler example: the equation $pq - qp = h$ passes into the classical $pq = qp$ when $h \to 0$.

The founders of QM wrote often about this limit transition, about the correspondence between the quantum and the classical magnitudes, etc. Nevertheless, none of them gave an explicit formulation of the CP.

It is easy to notice that the relation of Einstein's Special Theory of Relativity (STR) to CM is analogous to the relation of QM to CM in considered respect. When we assume that the light velocity c tends to infinity, many basic equations of STR pass asymptotically into the equations of CM.[2] E.g. the relativistic equation

$$m = \frac{m_0}{\sqrt{1 - \dfrac{v^2}{c^2}}}$$

passes into the classical one $m = m_0$.

A similar relation occurs between the General Theory of Relativity (GTR) and the classical theory of gravitation, etc.

When we grasp the CP in such a generalised way, it turns out that Einstein used it before Bohr, although he did not mention it explicitly. However, the physicists mostly apply the expression 'correspondence' only to the relation of QM to CM. They stress sometimes even an essential difference between STR and QM: the STR may be formulated autonomously, without mentioning the CM; in the case of the QM it is impossible (Bohm, 1951, p. 625).

Nevertheless, the analogy of the logical relations between the new and the old theory in both cases is very strong, which excuses the common expression *Correspondence Relation* (CR). Moreover, the CP has a wider value. As we shall see, it is a general principle of development of physics and, probably, of every advanced science. We can add that the analysis of these questions is a task of the philosophers of science.

1.2. THE ATTITUDE OF PHILOSOPHERS

Scientists seldom discuss methodological principles. The CP is one of few methodological principles formulated, though in a limited range and not in a clear form, by a physicist.[3] Another, although more controversial, principle is that of Complementarity, also formulated by Bohr. A third is Einstein's principle of relativity.

One would expect that the professional methodologists – the philosophers of science – will rush to this subject and discuss it animatedly, the more so since the CP is a principle of the development of science. But – contrary to the case of complementarity and of relativity – in the case of the CP philosophers were silent for a long time, they rarely noticed this principle at all!

The period of the 20's and 30's was not favorable to the CP. The development of science was not a domain philosophers were interested in.

The dominant place in the European philosophy of science was occupied by the neopositivists of the Vienna Circle and like-minded philosophers in other countries. They were interested in problems of the logical foundations of science, they analyzed questions of basic empirical statements, of the verifiability of scientific sentences, of analytical and synthetical sentences, of the syntax and semantics of scientific language, etc. (and the analysis of all these problems was a great merit of neopositivism). The main subject of their discussion was the problem of demarcation between science and metaphysics. They said much about the postulate of coherence inside a theory but not among different theories arising successively in science. In general, their attitude was ahistorical, the growth of science was outside their interests. In addition, they considered only very simple idealized models of science and payed no attention to real science and its problems.

The rationalist philosophers from the late Neo-Kantian and related streams were interested in the active role of the mind in knowledge, in the *a priori* elements in it, in the dialectic of object and subject, in the criticism of empiricism and positivism. They spoke sometimes about the development of knowledge but in a very abstract, speculative manner. They did not analyse real science and its development. Only sometimes they took examples from the history of science to illustrate their reasoning.

The Neo-Thomists and other religious philosophers payed even less attention to science, only rarely attempting to use some scientific discoveries to support their philosophical ideas.

Marxist philosophy was always interested in the history, including the development, of science. Engels in 1873–1882 (when he was working on *Dialectic of Nature*) and Lenin in 1908 (when he was writing *Materialism and Empirico-Criticism*) paid much attention to science and its history.

However, in the 20's and 30's in the U.S.S.R. the climate for the analysis of the CP and many other problems was not favorable. The increasing tendencies toward dogmatism hampered analysis of new achievements in science. The philosophers mostly limited themselves to a search for examples supporting some ideas of the classics of Marxism. There were also additional reasons. Soviet philosophers thought at that time that the CP was incompatible with Lenin's reflection theory, according to which a scientific theory should be confronted with reality (by means of practice, of experience) and not with another theory. The CP was held to be akin to the coherence conception of truth associated with subjective idealism (because it considered only relations among statements and not a relation between a statement and reality). Therefore, a common attitude toward the CP was reluctance, which is now recognized as a striking misunderstanding (cf. Kedrov, 1969b, p. 91). The situation was, indeed, paradoxical: the Vienna Circle and Soviet philosophy made the same mistake in postulating the confrontation of a theory only with experience (though differently interpreted) and not with other theories.

However, this situation did not last long. Soviet physicists spoke constantly about the CP. Philosophers arrived gradually at the conclusion that this principle is perfectly compatible with dialectical materialism, especially with the idea of the development of knowledge through relative truths towards the absolute one. In this case a crucial change could occur, even during the Stalin period. I.V. Kuznietsov, one of the leading Soviet philosophers of science, published a small but important book — the first book in world literature devoted especially to the CP (Kuznietsov, 1948). He appreciated very highly Bohr's merits and considered the CP as a general principle of the development of science, in any case of physics. He gave numerous examples of the correspondence relation — we shall cite them below (1.3).

Some Soviet philosophers greeted this book reluctantly. One of them wrote that, according to materialism, the criterion of truth is practice, experience, and Kuznietsov proposes an additional criterion — the correspondence between theories, which is a deviation from materialism (Shakhparonov, 1951). Nevertheless, in a short time most Soviet philosophers of science accepted CP as one of the basic laws of the development of science. One author criticised Shakhparonov for not understanding that 'developed' materialism is not primitive empiricism. At the same time he made a critical remark on Kuznietsov's thesis that the CP is 'a child' of twentieth century physics; he claimed that the idea of CP was contained in the concept of dialectical negation developed by Hegel and Engels long before Bohr (Arsieniev, 1958). I cannot agree: a general philosophical idea must not be identified with a much more concrete methodological principle.

During the next decades Marxist philosophers of science in the U.S.S.R. and other countries often mention the CP, usually citing Kuznietsov's book. However, they rarely add something new and do not analyse this principle more thoroughly. Some attempts in this direction are contained in a paper published in the D.D.R. (Laitko, 1969). Recently there appeared in Moscow a book devoted to the memory of I.V. Kuznietsov, who died in 1970. Five papers are concerned with the CP (Kedrov and Ovtchinnikov (eds.) 1974). They also do not notice logical difficulties connected with it.

The interest in the CP among western philosophers of science was much weaker for a long time. In the 50's this principle was noticed by some authors. Karl Popper mentions the CP in 1956 as an 'extremely fruitful principle' (see Popper, 1963, p. 101); in 1957 he defines it as 'a demand that a new theory should contain the old one approximately, for appropriate values of the parameters of the new theory' (see Popper, 1972, p. 202). Another author writes that the correspondence is 'an important idea of modern physics' giving some examples (Hutten, 1956, p. 101). At a Colloquium in Warsaw in 1961 some authors mentioned the CP, discussing the problem of reduction and justification of theories (Witt-Hansen, 1965; Groenwald, 1965). However, none of them analysed this principle more thoroughly.

In the 60's the interest in the development of science increased in western philosophy to a large extent. However, some authors active in this domain represent a relativistic approach to science and are reluctant toward the CP. T.S. Kuhn sees no correspondence among theories divided by a scientific revolution (Kuhn, 1962). P.K. Feyerabend goes even further and says that CP played a role in the early quantum theory but now it must be rejected because the *Korrespondenzdenken* hampers the revolution in science (Feyerabend, 1962b). Many philosophers of science polemicize about the relativism (we shall consider these problems in Ch. 5 and Ch. 7) but usually they do not mention the CP.

At the turn of 60's and 70's interest in the CP increased. Martin Strauss[4] constructed an Inter-Theory Relation theory (ITR) in which the correspondence relation (limit relation) occupies an important place (Strauss, 1970, 1973). In England Noretta Koertge wrote a Ph.D. thesis on the CP and continues this theme (Koertge, 1970, 1973). H.R. Post formulated a General Correspondence Principle considering it as one of the general laws of the development of science (Post, 1971). In the U.S.A. Aage Petersen wrote about the CP in QM (Petersen, 1968).

An analysis of the logical, methodological and historical aspects of the CP was made in Poland in the 70's. It was initiated by two Ph.D. theses written in Poznań by Izabella Nowak and in Łódź by Wiktor Niedźwiedzki and continued by other philosophers of science (see Krajewski *et al.* (eds.) 1974 and Kmita (ed.) 1974). We shall present its results below.

1.3. A GENERAL METHODOLOGICAL PRINCIPLE IN PHYSICS

We shall consider the CP as a general methodological principle of the development of an advanced science, and in any case a principle valid for contemporary physics. According to this principle, when we replace an old law by a new one containing some new parameters, the former is not eliminated; its equation is a limit case of the new equation under the condition that the new parameters have some extreme values (usually zero).[5] Then the new law (or, strictly speaking – new equation) passes asymptotically into the old one when considered parameters tend to the extreme value. When these parameters have values close to the extreme, the old law preserves its validity with a good approximation. We may apply the CP also to theories: a new theory T_2 is in a CR with an old one T_1 when there is a CR between basic laws of T_1 and T_2.

This relation already took place among some theories in the nineteenth century. For example: geometrical (ray) optics and wave optics; the latter passes into the former when the wavelength diminishes: $\lambda \to 0$. Another example (which will be examined more precisely in 2.2.): Boyle-Mariotte's law and van der Waals' law; the latter passes into the former when we may neglect the intermolecular forces ($a \to 0$) and the volume of molecules ($b \to 0$).

In twentieth century physics there is an abundance of examples. As we know, QM passes into CM when $h \to 0$, STR into CM when $c \to \infty$.[6] The relativistic quantum mechanics created by P. Dirac passes into QM when $c \to \infty$ and into STR when $h \to 0$; it passes into CM when both $c \to \infty$ and

$h \rightarrow 0$. GTR passes into STR when gravitational potentials $g_{ik} \rightarrow 0$ when $i \neq k$ and $g_{ik} \rightarrow 1$ when $i = k$. The Bose-Einstein and Fermi-Dirac statistics pass into the classical Maxwell-Boltzmann statistics when the temperature increases $T \rightarrow \infty$. The theory of nuclear forces which assumes its transfer by mesons μ passes into the Maxwell electrodynamics when $\mu \rightarrow 0$ (all these examples are contained in Kuznietsov, 1948). Also the newest theories fulfill the CP, e.g. the hypothetical quantum theories of space and time assume the existence of an elementary length l_0 and an elementary time-interval $\mathcal{T}_0$. These theories pass asymptotically into the non-quantum theories of space and time when $l_0 \rightarrow 0$ and $\mathcal{T}_0 \rightarrow 0$ (cf. Kuznietsov, 1968).

This regularity will probably persist in the future development of physics. The more precise its methods are, the more valid are the results. We may hope that in contemporary physics every law is established 'forever' and later development will only show its limits. This is held by some physicists to be the 'greatest achievement of science' (quoted after Kuznietsov, 1948). However, some qualifications must be added. The old law, as we shall see, must be reinterpreted in the light of a new one. Nevertheless, it persists in a new version, it is not eliminated. The rejection of old theories can be found only at the early stages of science, in the period of its pre-maturity (see Ch. 7).

Analogous relations may be found in mathematics. The geometries of Lobatchevsky and Riemann are generalizations of Euclidean geometry. The latter is a limit case of both of them when the curvature radius $K \rightarrow \infty$. There are analogous relations among the successive concepts of number, e.g. the set of complex numbers $a + bi$ passes into the set of real ones when $b = 0$. Citing these and similar examples many Soviet authors say that the CP is fulfilled also in mathematics (cf. Kagan, 1944; Nyssanbajev, 1965, 1974; Kuznietsov, 1970b). This may be true, but I prefer to confine the analysis to physics — there are logical problems here which do not arise in mathematics.

Let us notice that some authors speak about similar relations (without the expression 'CP') between science and common sense, e.g., logical concepts of truth, meaning, proof, etc. must not be defined arbitrarily: they must be in some 'limit cases' in accordance with the 'naive' view (Popovitch, 1969). To be sure, the concept of limit does not have an exact mathematical meaning in this case.

1.4 DESCRIPTIVE AND NORMATIVE VERSIONS

Like many other methodological principles, the CP may be interpreted in two ways which may be called *descriptive* and *prescriptive* (cf. Meyer-Abich, 1965, p. 87) or *descriptive* and *normative*. In the first version we have a description of a relation which (always or often) takes place between successive theories in science. In the second version we have a norm which ought to be fulfilled when one creates a new theory.

Both versions are closely connected. True, a norm does not follow logically from a description. Nevertheless, the more often a relation is met in real science (especially –in advanced science) in the past, the more probable is its occurrence in the future, which is a hint that it should be respected in scientific work if one believes in the regularity of the development of science.

Of course, a scientist may also formulate a norm independently of its validity in the past. In other words, he may postulate a norm and not discover its use, as historians do. This was the case of Bohr, who did not examine the history of physics but invented a methodological norm for his own aim (probably not noticing that this norm in a generalized way governed the development of physics for a long time). Bohr introduced his principle because he had great difficulties with the experimental proof of his new theory of the atom. Otherwise, he probably would not have done so. In general, scientists begin to reflect upon methodology only when something is 'wrong'. We see that methodology has profited from difficulties in science. True, if Bohr had not invented the CP, it probably would have been discovered sooner or later by other scientists or by philosophers interested in problems of the development of science.

Einstein met difficulties of another kind than Bohr did, but nevertheless he was also careful to show the accordance, in limit cases, of both STR and GTR with CM. Hence, he used in fact the methodological version of the CP. When Einstein created GTR he proved at first that the Newton (or Poisson) equations are the first approximations of his equations; afterwards he predicted deviation in some cases – e.g. that of Mercury's movement – of the values obtained by means of his theory from the values obtained by means of Newton's. As is known, the observations have yielded positive results.

In general, a new theory is at first confronted with an old one and only after passing the proof of correspondence is it confronted with experiments. A theoretical proof precedes an empirical one. Notice also that these two proofs have opposite 'directions': the first — towards the accordance with the old theory, the second — towards the difference. The theoretical proof is 'for' the old theory, the empirical one 'against' it (cf. Such, 1974).

Philosophers of science are in most cases interested in the descriptive version of a methodological principle. They try to explicate it and examine it from a logical point of view (when they have logical inclinations). They search for the validity of this principle in different branches of science, at different stages of its development, etc. (when they have historical inclinations). However, many philosophers are interested also in the normative version. Some of them formulate or explicate a norm in order to draw scientists' attention to it, to 'teach' them (when they have purely methodological and 'didactic' inclinations). Others investigate the functioning of a norm in science, the degree to which it is realized by scientists admitting it, the motives of its use, the conditions which foster or hinder it, etc. (when they have psychological and sociological inclinations). In the last case they describe the functioning of a norm, and we may speak about the descriptive approach to the normative version of a methodological principle.

Logical analysis of the descriptive version of the CP will be the subject of the next sections and chapters. Now let us briefly consider the normative version. It can serve (and it did in Bohr's work) two functions: (1) an *instrumental* one — as an instrument ('a magic stick') for the search for new laws, (2) a *justifying* one — as a way of justification of new ideas admitted arbitrarily (e.g. quantum postulates) (cf. Lewenstam, 1974). The CP in the normative version prevents the loss of achievements of previous stages of science. According to this principle, the domain of T_1 must be described by T_2 at least as well as by T_1 (Rubinowicz, 1961). The CP resembles in this respect the Hippocratic maxim in medicine: *primum — non nocere* (Szumilewicz, 1974). The CP prevents arbitrary speculations. It resembles Fr. Bacon's remark that the mind needs lead (Stakhanov, 1974). Sometimes one notices a psychological role of the CP: it mitigates the resistance which a new idea usually meets within the scientific community (Pietruska-Madej, 1974).

1.5. SOME LOGICAL DIFFICULTIES

When we consider successive physical laws or theories formally, syntactically, i.e. when we consider mathematical equations, then the correspondence relation is clear. When a parameter tends to an extreme value, then the new equation passes asymptotically into the old one. The situation changes when we pass from this syntactical approach to the semantics, i.e. when we consider the physical sense of the parameters. The characteristic for a new theory parameter in fact never takes on the extreme value — it is physically impossible.

In some cases this parameter may approximate the extreme value. Take the relation between geometrical and wave optics. We may consider here radiation with lesser and lesser wavelengths. Thus the assumption $\lambda = 0$ is approximately fulfilled (not exactly, of course: the wavelength cannot take on arbitrarily low values). In the case of the van der Waals law the situation is similar. We may consider different gases in different conditions and diminish a and b though not to arbitrarily low values.

In the cases of QM, STR and many others the situation is much worse. Quantum action h and light velocity c are absolute constants. Their constancy is a theoretical postulate, their values are known from experience: $h = 6.22 \times 10^{-27}$ erg·sec, $c = 3 \times 10^{10}$ cm sec^{-1}. Since h and c are constant, they cannot change and the assumptions that they tend toward a limit is notoriously false.

Nevertheless, in the limit transition from QM or STR to CM we make these assumptions. The situation is paradoxical. We draw conclusions from false assumptions although we know from elementary logic that 'the falsity entails everything'. Hence, we perform a kind of reasoning forbidden by logic.

Some physicists realise this and say that we must understand the assumptions about the change of h (or c) 'conventionally' (e.g. Davydov, 1963). However, this is no solution of the issue, only a hint that it is noticed. Maybe physicists can limit themselves to this. However, philosophers cannot. Nevertheless, philosophers of science have been hitherto unmoved by this strange situation, they rarely even notice it at all! Neither I.V. Kuznietsov nor his numerous followers noted this paradox.

Recently some authors have seen this difficulty. M. Strauss considers the limit transition from STR to CM when $v/c \rightarrow 0$ and distinguishes two possibilities: (1) $v \rightarrow 0$, (2) $c \rightarrow \infty$. The operator (1) is defined within STR but it is uninteresting (only a special value of v). The operator (2) is forbidden by STR; it may be defined in a metatheory which treats c as variable (Strauss, 1970). But what is this metatheory? A sequence of STRs describing a sequence of worlds with different values of c? This belongs rather to a domain of science fiction.

There is also another difficulty. The new theory casts a new light on the old one. Take the relation of STR to CM. We interpret the laws of CM in the light of STR otherwise. We know that they are only approximately valid when $v \ll c$. Some theses of STR are incompatible with CM (e.g. dependence of mass on the coordinate system). How then may we say that we infer the laws of CM from the laws of STR and some limit conditions?

We can add that analogous difficulties have appeared in classical physics since the time of Newton. Already P. Duhem noticed that Keplerian laws are, strictly speaking, incompatible with Newtonian ones and we may deduce from the latter only approximations of the former (Duhem, 1906).

Now many philosophers say that old, empirical laws demand a 'correction from above' in the light of new, deeper theories (Feigl, 1964), that a new theory explains an old one, correcting it (Tuomela, 1973), that a new theory generalizes an old one and corrects it (Krüger, 1973), that a new theory entails a qualified version of the old one (Koertge, 1973). However, they usually do not analyze the nature of these corrections or qualifications.

Kuhn and Feyerabend stress these difficulties but draw from them nihilistic conclusions: they reject the CP altogether. S. Amsterdamski accepts the syntactical version of the correspondence relation but rejects the semantical one because of 'meaning variance' (Amsterdamski, 1973). We shall discuss the problem of 'meaning variance' in Ch. 5. In any case, we will maintain the CP also in the semantical version. However, the logical difficulties described above must be solved. In order to solve them, we must analyze the nature of scientific laws and the relations among them. We start with an analysis of the processes of idealization and factualization that play, as it seems, a crucial role in this case.

NOTES TO CHAPTER 1

[1] An author notices that the classical limit is approached when (1) $h \rightarrow 0$, (2) the mass becomes large, (3) wave-lengths become small, (4) dimensions become large, (5) quantum numbers become large (Lassile, 1971).

[2] To be sure, not all of them. The equation $E = mc^2$ does not pass into any equation of CM.

[3] Lately some physicists formulate the CP in a clear way, as a general methodological principle (Rubinowicz, 1961).

[4] Strauss is a German 'philosophying' physicist. He had been publishing before the war in *Erkenntnis*. At present he is working in the D.D.R. publishing there and in the West.

Hence, he associates the traditions of West and East.

[5] Sometimes — the infinity; but then we may take an inverse parameter, e.g. $1/c$ instead of c. Therefore, we may assume that the parameter tends to zero without limiting a generality of our reasoning.

[6] We use here, for short, the expressions 'STR passes into CM', etc., although, as we have noticed, not all equations of the latter pass into equations of the former.

CHAPTER 2

IDEALIZATION AND FACTUALIZATION

2.1. SCIENTIFIC LAW AS AN IMPLICATION

Let us start with the concept of law. To put it briefly: we distinguish laws of Nature i.e. constant relations in reality, and scientific laws, i.e. sentences stating these relations. Sometimes these are called laws$_1$ and laws$_2$ (cf. Bunge, 1959, 1967). This distinction is important in the general theory of knowledge (as a manifestation of epistemological realism), however, not in the methodology of science. There is no direct, non-verbal way of penetration into reality (as intuitionists claim), hence we can speak in practice only about law sentences formulated in the sciences. We shall call them simply *laws*.

Statements treated as candidates for laws, i.e. law-like statements, are always strictly universal ones, i.e. they must not contain proper names, space-time coordinates, etc. This is now commonly acknowledged among philosophers of science; some of them, however, say that it is not enough. Many Marxist philosophers claim that a law is not only a universal statement but it is a necessary one too. However, they do not explain what should be added to the universality. I think that we may identify real necessity with strict universality (cf. Krajewski, 1968).

Philosophers of science often say that a law is a *true* law-like statement. However, usually it turns out that a law formulated in the sciences is not strictly true. Therefore, it is better to say that we treat as laws these law-like statements which are considered by scientists to be true or, at least, approximately true (see Ch. 8).

The relations stated by laws are of a different nature, hence there are different kinds of laws: qualitative and quantitative, synchronic (coexistential) and diachronic (stating causal relations, stages of development, etc.), unique (deterministic) and statistical, etc. We may combine these divisions and obtain a complex classification of laws (Krajewski, 1972a). In this book we shall consider mainly quantitative synchronic deterministic laws. They state usually a functional dependence F between some parameters $a, b, \ldots$

of each object (of a certain kind) x. We shall write it as an equation:

(2.1.1) $F[a(x), b(x) \ldots] = 0$

or in a shorter form

(2.1.1') $F(x) = 0$.

Scientists usually consider this equation to be a law. However, it is fulfilled only under certain conditions C. In textbooks of physics the description of C is usually treated as a commentary on the law. This does not suffice: C is an integral part of the law. There are different ways to express this logically. One way is to use a model theory: one considers a model M in which C is realised and $F(x) = 0$ is true. Many logicians choose this way now. For our purposes another one will be more convenient: the use of implication preceded by a general quantifier. The general scheme of a quantitative law may be presented as follows:

(2.1.2) $\bigwedge_x [C(x) \Rightarrow F(x) = 0]$.

Here x is a material object, $C(x)$ a set of conditions (more precisely: a conjunction of statements describing the conditions), $F(x) = 0$ a dependence (equation) fulfilled under these conditions.

Physicists regard as a law the equation $F(x) = 0$ and say that it is fulfilled under some conditions. Also logicians using the model theory regard the equation as a law and say that it is true in some models. We shall regard the whole implication (2.1.2) as a law. It is always fulfilled: when $C(x)$ is true the law is non-emptily fulfilled, when $C(x)$ is false the law is emptily fulfilled. We may say that in this form the law, if true, is eternal (cf. Krajewski, 1973a, Ch. VI).

In the course of the development of science both C and F change. When F_1 is replaced by F_2 physicists say that we are dealing with a new law. When, however, C_1 is replaced by C_2, F being unchanged, they say that we have to do with the same law. It is obviously so when $F(x) = 0$ is treated as the law. In our treatment the law changes in both cases. Nevertheless, in order to be possibly close to the terminology used by scientists, we shall say that when F changes we have a new law and when C changes we have a new expression (formulation) of the same law (cf. Mejbaum, 1964).

C determines a set of objects which we shall call a *domain* of a law and designate by D. C formulated in a given period of science determines the *supposed domain D* in that period. Usually in the course of the development of science D becomes narrower because new conditions limiting the validity of $F(x) = 0$ are revealed. We may speak about a temporal sequence of expressions of a law when $F(x)$ remains unchanged and we formulate successively conditions $C_1, C_2, C_3 \ldots$ and obtain respective domains $D_1, D_2, D_3 \ldots$ We may also speak about the genuine domain D_0 of the law in which $F(x) = 0$ is indeed true (absolutely true). Of course, we never know whether we have found it. It is probably an unattainable limit of the sequence D_i:

$$(2.1.3) \quad D_0 = \lim_{i \to \infty} D_i.$$

Consider now changes of F. Usually a function F_1 is replaced by a more complex function F_2 (with additional terms, etc.). We have a new law then. However, it is a new law_2 which reflects the same law_1. We have a temporal sequence of $laws_2$ which approximate more and more the exact description of the law_1. In other words, the limit of a sequence $F_1, F_2, F_3 \ldots$ is the genuine dependence F_0:

$$(2.1.4) \quad F_0(x) = \lim_{j \to \infty} F_j(x).$$

Hence, we have three concepts: a real law_1 of nature, a scientific law_2 and its expression. We may call them a 'conjugated triple' (cf. Such, 1972, Ch. IV).

2.2. FACTUAL AND IDEALIZATIONAL LAWS

When the supposed domain of a law is a set of real, empirically existing objects, we call this law *factual* (L_F). When we know that a law has no real domain, we call this law idealizational (L_I). The *idealizational* law has, of course, also a domain. It is, however, not a fragment of reality but a set of *ideal objects* (models, constructs). In the first case the domain is real, in the second one ideal.

We speak about the supposed domain. When the supposed domain is ideal, probably the genuine domain D_0 is also ideal. When, however, the supposed domain is real, we do not know whether D_0 is also real. It has happened often in the history of science that a law at first considered to be factual turned out to be idealizational afterwards. We shall give two examples.

R. Boyle (strictly speaking, his disciple R. Townley) discovered that the volume of a given mass of gas is inversely proportional to the pressure: $pV = $ const. E. Mariotte discovered that this dependence is fulfilled only when the temperature of the gas is constant. We may say that Boyle discovered the equation $F(x) = 0$ and Mariotte stated the conditions C and the whole implication.[1] However, both were convinced that the law is factual. This conviction persisted two centuries. Only the molecular-kinetic theory of gases revealed that this law is valid for *ideal gas*. For *real* gas another law is valid, the one discovered by van der Waals:

$$\left(p + \frac{a}{V^2}\right)(V - b) = \text{const.,}$$

where a characterizes the average intermolecular force, and b the volume of molecules. In ideal gas, molecules are material points ($b = 0$) and there are no intermolecular forces ($a = 0$). When we suppose in the van der Waals equation $a = 0$ and $b = 0$, we obtain the Boyle equation. It is a classical example of correspondence relation (limit transition) in nineteenth century physics.

We know now that the Boyle-Mariotte law is an idealizational law. It is exactly fulfilled in ideal gas and only approximately in real gases in which a and b are low (e.g. in hydrogen in low pressures and temperature far from the critical one). Is the van der Waals law a factual one? It yields a much better approximation to the behaviour of real gases, nevertheless also only an approximation. Van der Waals' law neglects some other less important factors which disturb the dependence. Therefore, it is also an idealizational law but of a 'lower degree'. It is 'more factual' than the Boyle-Mariotte law.

Another example: G. Ohm discovered that when electric current flows through a conductor, the tension at its ends equals the product of the intensity of current and the resistance of the conductor: $V = IR$. He believed that it is a factual law. However, it soon turned out that this law is not exact. When the circuit contains a condenser with electric charge Q and capacity C and a coil with self-induction L, another law is valid, the so-called differential law of current. It has the following form:

$$RI + L\frac{dI}{dt} + \frac{Q}{C} = V.$$

When $L \to 0$ and $C \to \infty$ we pass to the simple Ohm law. It is another classic example of the correspondence relation. Each conductor has some finite L and C, even where there are no coils and condensers in the circuit. Only in the *ideal conductor* $L = 0$ and $C = \infty$. Hence, the Ohm law is idealizational. The differential law of current is more factual.

We see that two elementary laws of physics are idealizational. Their domains are ideal models (ideal gas, ideal conductor). In other words, these laws are *counter-factual*. Their antecedents C are never fulfilled in reality. They are true only in ideal models. Hence, we must introduce a concept of a model truth (relation of a sentence to the model) apart from the classical concept of truth as a relation to reality (see Ch. 8). We shall distinguish factual statements C_F, which are true in a classical sense, and idealizational statements (idealizing assumptions) C_I which are true in a model sense only.

Although idealizational laws are counterfactual, they have an empirical content, so they can be and must be empirically tested, usually after a procedure which may be called factualization. We shall speak about it in (2.5).

2.3. IDEALIZATION IN SCIENCE

The method of idealization is one of the main methods of every advanced science. Probably, the first who used this method consciously was Erathostenes, measuring the diameter of the Earth. He assumed that the Sun's rays in two distant places are parallel, i.e. that the distance of the Sun is infinite, although he knew, of course, that this assumption was false. Here the method of idealization was used not yet for the discovery of a law but for the discovery of a fact unattainable by direct experience — the size of the Earth (cf. Such, 1973).

The first idealizational law of physics was, probably, Archimedes' law, according to which a lever persists in an equilibrium when the moments of forces acting on it (the products of forces and their arms) are equal. Archimedes assumed that the lever is weightless although he realised that such levers do not exist. Hence, statics was the first branch of physics in which idealizational laws were formulated.

Ancient dynamics did not reach the stage of idealization. Aristotle claimed that (apart from 'natural movements') a body is moving when a force (a motor) acts on it and stops moving when the force ceases. This statement was a simple generalization of everyday experience and we may describe this law as factual. Aristotle did not want to consider counterfactual circumstances, e.g. the vacuum which he supposed did not exist. We may see how delusive the common opinion is that ancient science was speculative, not empirical. It had, of course, speculative elements; nevertheless, its main weakness was its crude empiricism. In astronomy crude empiricism led the ancient philosophers to geocentrism. Copernicus and Galileo had to conquer not only scholastic dogmatism but also the crude empiricism of ancient science.

Galileo was the master of idealization. His law of inertia is a classic example of an idealizational law. It assumes following idealizing conditions: the vacuum; the absence of friction and of any other forces acting on the considered body. Galileo's law of free fall assumes also many idealizing conditions: the vacuum, the absence of any forces besides the Earth's gravity, the infiniteness of Earth's radius (only then the acceleration g remains unchanged during the falling). Galileo was aware of the idealizational nature of these laws, though he did not use the term 'idealization' (he spoke about 'abstraction'). In his *Dialogue* he criticizes from this point of view Aristotle, 'who preferred the sense experience to any reasoning', and extolled many times the abstract, mathematical reasoning. He praised Aristarchus and Copernicus for the audacity of their claims incompatible with eye witnesses, etc. Galileo usually considers ideal spheres, perfectly smooth and horizontal planes, etc. In the *Dialogue* Simplicio (an adherent of Aristotle) often says that such ideal objects do not exist; Salviati (porte-parole of Galileo) explains to him the importance of considering things in their pure form, of temporarily disregarding accidental factors, etc. (Galileo, 1638). For reason of his 'mathematicism' Galileo is often described as a platonist who fought against aristotelianism. However, he claimed — contrary to Plato — that mathematical construction concerning Nature ought to be tested by experience. Galileo was a true disciple of Archimedes (cf. Shea, 1972). Therefore, we may dub his methodology *archimedianism* (cf. Krajewski, 1975).

After Galileo the method of idealization entered firmly into physics. It was used, of course, by I. Newton. Such basic concepts of CM as material point, inertial system, etc. are ideal models. However, Newton was an inductivist and he did not realise the use of idealization so clearly as Galileo did (cf. Such, 1972b). Later many ideal models were introduced into physics: ideal gas, perfectly elastic body, absolutely black body, ideal thermodynamic

processes (isothermic, isobaric, adiabat), etc. All laws containing these concepts are, of course, idealizational.

The same method penetrated into other sciences, including the social
sciences. In economy the method of idealization was used by D. Ricardo
and to a large extent by K. Marx. In the first volume of *Das Kapital* Marx
considers a pure capitalism (in which there are only two classes: capitalists
and workers), a closed national economy (without foreign trade) and makes
many other simplifying assumptions.[2] He often notices explicitly that his
assumptions contradict the facts.

The simplest example of Marxian idealizational laws is the value law
stating that the price of an article is determined by its value, i.e. the amount
of work necessary (in given technical conditions) for producing it. In reality
prices deviate from this value because of fluctuations of supply and demand.
However, the law of supply and demand does not explain the price. It is
obvious when supply and demand are equal. Then, says Marx, we must
search for other causes as is done in physics where, in the case of a balance
of two opposite forces, the movement must be explained by other causes
than the two balanced forces. Marx criticizes 'vulgar economists' for appeal-
ing to direct experience and not understanding the proper method of science
which calls for considering phenomena in their 'pure form' not disturbed by
'secondary factors' (Marx, 1867). The analogy with Galileo is so striking
that Marx may be called 'Galileo of the social sciences' (Nowak, 1971b).

In modern economy mathematical models are widely used and they are
always idealizations. The situation is similar in modern biology, modern
linguistics and in other advanced sciences. Some scientists understand it
more, others less clearly. There is nothing strange about it. The methodological status of science is rarely analyzed by scientists themselves. Its
analysis is a task for philosophers.

2.4. THE ATTITUDE OF PHILOSOPHERS

The situation is not favorable. Philosophers of the nineteenth and twentieth
centuries mostly do not notice the crucial role of idealization in science.
Many of them do not mention it at all.

In the nineteenth century the role of idealization in science was understood by Marx who was, as we know, not only an economist but also a

philosopher. He used (like Galileo) the word 'abstraction'. He wrote that in economy it is necessary to start from abstract and the simplest concepts and to 'climb' gradually to the 'concrete' (*aufsteigen vom Abstrakten zum Konkreten*). The 'concrete' is the starting point in an analysis of empirical material but it is the final result in a theoretical reasoning. The method of 'climbing from abstract' is the sole method of 'assimilation' of the 'concrete' by the thinking mind (Marx, 1859).

'Abstract' means here 'ideal model'. Therefore, the Marxian method of climbing from abstract to concrete is now called by the Poznań methodological school in Poland the 'method of idealization and gradual concretization'. I prefer to call the latter 'factualization'.[3]

Friedrich Engels noticed the great role of idealization in physics when he considered the work of Sadi Carnot. His 'ideal steam-engine' describes the process of transforming heat into movement in a 'pure, independent, untainted form' (Engels, 1925). However, neither Marx nor Engels developed this theme.

The main trend in the methodology of science in the nineteenth and in the first half of the twentieth century was radical empiricism (positivism). It considered facts (sometimes interpreted subjectively, in other cases objectively) to be the sole reality and the absolute basis of the science. It considered laws of science to be simple generalizations of facts by means of induction (inductivism). In such a simplified image of science there is, of course, no place for idealization. K. Popper criticised the inductivism and developed a program of hypotheticism (see Ch. 5). However, he also does not note the crucial role of idealization in science.

Another current in philosophy is rationalism. German Neo-Kantians at the end of the nineteenth and the beginning of the twentieth centuries spoke much about the idealness (*Idealität*) of the object of science (cf. Natorp, 1910). However, they were apriorists, they regarded the whole object of science as an ideal one, including facts, and did not distinguish idealizational and factual laws. French rationalists at the beginning of the twentieth century were closer to real science. They said much about ideal constructions in science, about the counterfactual nature of laws of physics, but they also admitted the existence of aprioristic schemes in science (Meyerson, 1908, 1931; Metzger, 1926). We meet many interesting antipositivistic ideas in papers by conventionalists (Poincaré, 1902, 1906;

Duhem, 1906) but none of them distinguished idealizational and factual laws.

In the second half of the twentieth century the whole western analytical philosophy of science gradually overcomes its older positivistic narrowness. It begins to appreciate the role of theoretical thinking and theoretical models in the empirical sciences. But the role of idealization is rarely noticed. Some authors connect it with causality. They say that 'strictly causal laws' hold only among 'ideal physical states' (Reichenbach, 1946, p. 2), that causality is a relation in the sphere of 'conceptual events' or 'ideal models' (Lenzen, 1954, p. 6). However, not only 'strictly causal laws' have the idealizational nature. We saw with the example of gas laws that statistical laws of theoretical physics do not differ in this respect from strictly deterministic ones.

Many empiricist philosophers of science consider the problem of theoretical 'constructs' and call them sometimes 'ideal objects' in contradistinction to empirical objects. However, the concept of construct is ambiguous. Atoms, electrons, photons, etc., often held to be constructs, really exist, although they are not given in direct experience and a theory is necessary for stating their existence. Ideal models (material points, inertial systems, ideal gases, absolutely black bodies, etc.) do not exist in reality. Perhaps the existence of electrons is conjectural but the non-existence of material points and other ideal objects is absolutely sure.

The empiricist philosophers of science usually take into account only factual laws; they mostly give simple examples from everyday experience or from elementary parts of science. Only in the last decades did some authors notice that science formulates idealizational laws too. They often claim that such laws are admitted only when they are deducible from more general factual laws (Nagel, 1961, p. 60; Hempel, 1965, p. 168). However, this claim is untenable — the general laws are also idealizational, they cannot be inferred from factual laws (cf. Kmita, 1972; Nowak, 1974b).

True, there is one law in which idealization is explicitly expressed — the law of inertia. Many philosophers of science notice that this law is counterfactual, but they regard it as an extraordinary exception which causes trouble for methodologists (e.g. Hanson, 1965). They overlook the fact that it is a typical situation.

Paradoxically, the role of idealization is better understood by many philosophers of social science. Max Weber wrote much about *ideal types* but he claimed that its use is a peculiarity of the humanities (Weber, 1922). A similar view is held by some contemporary methodologists of social science (cf. Rudner, 1966). Even Hempel writes more about idealization in the social sciences than in the natural sciences (see Hempel, 1965). This situation is paradoxical because, as we have seen, the idealization method has played a crucial role in physics since Galileo and other sciences only gradually approach physics in this respect.

Only in the last period some analytical philosophers have stressed the role of idealization in science (Humphreys, 1968; Achinstein, 1971). For others it is surprising. Even W. Barr, who analyzes the concept of idealization himself (Barr, 1971, 1974), wonders why Humphreys 'goes so far' as to state that the idealizational law is 'more like the norm of physical law' than the exception (Barr, 1971, p. 258).

However, even philosophers who notice the role of idealization usually do not consider the process of factualization. True, E. Nagel mentions the fact that some additional assumptions must be introduced to 'bridge the gap between the ideal case . . . and the concrete circumstances . . .' (Nagel, 1961, p. 463). But this is only a marginal remark.

In Marxist philosophy the role of idealization is often noticed. Many philosophers of science in the U.S.S.R. and in other countries speak about the introduction of idealizational concepts and laws as an important method of physics and other sciences. Some of them show the peculiarity of idealization as a special kind of abstraction where some magnitudes take extreme and fictitious values (Gorski, 1961; Subbotin, 1964). However, they do not analyze more thoroughly the logical form of idealizational laws and do not consider at all the process of factualization.

This task was undertaken in Poland, especially by the Poznań methodological school.[4] Its representatives, mainly L. Nowak in his numerous papers, consider the method of idealization and gradual concretization, primo, as the basic feature of Marxian methodology, secundo, as the method of all advanced sciences. Nowak's papers raised a lively discussion in Poland. Some authors criticised them for logical inaccuracy, others said that Nowak does not understand 'the genuine essence of Marxism', neglects its most essential features, etc. I think that Marxism may be differently interpreted and a discussion about the 'genuine essence' of Marxism is fruitless. It is not very important whether we call the considered method Marxian, or Archimedean, or Galilean, or otherwise. Only the second point is important for methodology – the role of this method in science. I think that Nowak properly grasped the main method of advanced sciences though he did not avoid some exaggerations (see 2.7).

2.5. IDEALIZATION AND FACTUALIZATION

We shall now consider the method of idealization and factualization more carefully.

As we have noticed, all advanced sciences deal with ideal objects, i.e. ideal models of real objects. Ideal objects have factual properties and ideal properties. The latter are so constructed that some parameters take on extreme values, usually zero. For example, the material point has a factual property, mass ($m > 0$), and an ideal property, absence of volume ($V = 0$).[5] Each real body has, of course, a finite volume ($V > 0$) but we may consider a sequence of bodies with diminishing volumes; an ideal limit of this sequence is the material point.

Generally speaking, x is an ideal object (or ideal model of real objects) when it is an ideal limit[6] of a sequence of real objects with diminishing value of some characteristic parameters p_i (usually, there is a set of parameters). In other words, some of its parameters are equal to zero ($p_i = 0$), however we know that in all real objects they are positive ($p_i > 0$).[7]

An idealizational law assumes two kinds of conditions: factual and ideal. We designate the set (the conjunction) of factual conditions by C_F and the set of ideal conditions, or *idealizing assumptions*, by C_I.

(2.5.1) $\quad C_I(x)$: $\quad p_1(x) = 0 \wedge p_2(x) = 0 \wedge \ldots \wedge p_n(x) = 0$.

Then we may write the scheme of an idealizational law L_I in a short form as follows:

(2.5.2) $\quad L_I$: $\quad \bigwedge_x [C_F(x) \wedge C_I(x) \Rightarrow F(x) = 0]$.

In a more developed, and more convenient for our purposes, form (hereafter we shall omit the general quantifier for brevity's sake):

(2.5.3) $\quad L_I$: $\quad C_F(x) \wedge p_1(x) = 0 \wedge p_2(x) = 0 \wedge \ldots \wedge p_n(x) = 0 \Rightarrow F(x) = 0$.

The ideal law is fulfilled only in the ideal models, hence it is not directly testable. For making it directly empirically testable we must transform it. We call this transformation a *factualization*. In order to factualize an idealizational law, we abrogate in turn the idealizing assumptions $p_i(x) = 0$ replacing them by the factual assumptions $p_i(x) \geqslant 0$.[8] We must then change also the functional dependence $F(x)$, and usually complicate it by introducing additional terms. We obtain successively idealizational laws of lower degrees (less idealizational or more factual) L_I', L_I'' ... and finally the factual law L_F:

(2.5.4) L_I: $C_F(x) \wedge p_1(x) = 0 \wedge p_2(x) = 0 \wedge \ldots \wedge p_n(x) = 0 \Rightarrow F(x) = 0$

(2.5.5) L_I': $C_F(x) \wedge p_1(x) \geqslant 0 \wedge p_2(x) = 0 \wedge \ldots \wedge p_n(x) = 0 \Rightarrow F'(x) = 0$

(2.5.6) L_I'': $C_F(x) \wedge p_1(x) \geqslant 0 \wedge p_2(x) \geqslant 0 \wedge \ldots \wedge p_n(x) = 0 \Rightarrow F''(x) = 0$

. . . .

(2.5.7) L_F: $C_F(x) \wedge p_1(x) \geqslant 0 \wedge p_2(x) \geqslant 0 \wedge \ldots \wedge p_n(x) \geqslant 0 \Rightarrow F^{(n)}(x) = 0$

The conditions $p_i \geqslant 0$ may be omitted (the next law is always more general than the preceding one) and we may write, e.g. in the last case:

(2.5.8) L_F: $C_F(x) \Rightarrow F^{(n)}(x) = 0$.

Nevertheless an exact factualization is rarely possible. Usually we do not know the exact equation for real objects. Probably each of our equations assumes some idealizing conditions although we often have no idea about them; only a further development of science reveals the assumptions that we must make. In any case, we usually must content ourselves with an approximate factualization, with an approximate factual law:

(2.5.9) L_F: $C_F(x) \Rightarrow F^{(n)}(x) \approx 0$.

We shall illustrate this procedure on the example of transition from the Boyle-Mariotte law to the van der Waals law. In this case:

$C_F(x)$:   is a mass of gas at constant temperature

$C_I(x)$: $a = 0 \wedge b = 0$.

We have a following sequence of laws (with decreasing degrees of idealization):

L_I: $C_F(x) \wedge a = 0 \wedge b = 0 \Rightarrow pV = \text{const.}$

L_I': $C_F(x) \wedge b = 0 \Rightarrow \left(p + \dfrac{a}{V^2}\right) V = \text{const.}$

L_F: $C_F(x) \Rightarrow \left(p + \dfrac{a}{V^2}\right)(V - b) = \text{const.}$

We presented here the van der Waals law as if it were a factual one. However, we know that it is fulfilled in real gases only approximately; it is also an idealizational law but 'two degree lower' than the Boyle-Mariotte law. The formulation of idealizing conditions is more complicated in this case. In technical physics there are some different, more precise (and more complicated) laws, e.g. the virial equation:

$$pV = A + Bp + Cp^2 + Dp^3.$$

The coefficients A, B, C, D are different for different masses, temperatures, etc. Their values are, of course, approximate, hence also the virial equation is approximately fulfilled in real systems (and we should write $\approx$ rather than $=$). It is, however, the best approximation and we may consider it as the final approximate factualization of the Boyle-Mariotte law.

It is remarkable that physicists do not usually call the virial equation a 'law' (cf. Achinstein, 1971, p. 13). The situation is paradoxical: physicists consider the equation fulfilled only in ideal models as law, and the equation fulfilled in real systems not. Achinstein sees the reason for this in the complexity of the virial equation although he admits that sometimes the physicists call complex equations 'laws'. I think there are other reasons. First, an idealizational law is fulfilled (in an ideal model) *exactly* and a factual law is fulfilled (in a real system) only *approximately*. Second, an idealizational law (but not a factual one) is included in a theoretical system; it may be inferred from other, more general laws. The Boyle-Mariotte law and even the van der Waals law are exactly fulfilled in some ideal models and may be inferred from statistical mechanics and some additional assumptions. The virial equation is approximately fulfilled in real gases and cannot be inferred from a general theory. We see here a deep difference in attitude between a theoretical science and a practical one.

2.6 IDEALIZATION AND ESSENCE

An important and difficult problem is the problem of the ontological status of ideal models and idealizational laws.

The positivists hold theories merely to be a convenient way for description and prediction of facts. The realists, especially the materialists, reject such instrumentalism and say that a theory reflects reality deeper then purely empirical laws. The direct experience shows us the 'surface' of the material world; theoretical science penetrates through this surface into the essence of the world. At the same time materialism radically differs from idealistic essentionalist philosophic systems (Thomism, phenomenology, Bergsonianism) which claim that the essence may be grasped only in a special 'philosophic' way. According to materialism, science and only science enables us to learn the essence of the world.

The situation is relatively easy when we deal with factual theories, e.g. theories about the internal structure of atoms, molecules, chromosomes, etc. Here the structure of a system is its essence. However, idealizational laws are counterfactual — here the situation is more difficult. Can a counterfactual statement tell us something about reality? I think so. Moreover, an idealizational law may reflect the essence of a real process.

We should distinguish in each real process the main (primary) factors and the side (secondary) factors. An idealizational law takes into account only the main factors. When we factualize it, we take the side factors into account one by one. We shall illustrate this with two examples already considered above.

Galileo's free fall law describes free falling, i.e. the movement of a body on which only the Earth's gravity acts. It disregards other factors (air resistance, wind, etc.) which disturb the fall in each real case. Gravity is the basic, essential factor of falling: if it did not act the phenomenon would not exist. Other factors are secondary because the fall is possible without them (of course, it is theoretically possible; practically their complete elimination is impossible). This hierarchy is objective: we may not regard air resistance as the primary factor and gravity as the secondary factor of falling. In this sense we may say that Galileo's law grasps the essence of falling.

Marxian analysis of capitalism has an analogous character. Capitalism is impossible without two basic classes constituting it: the capitalists and the proletarians. However, it is possible without other classes and social groups: peasants, artisans, intelligentsia, etc. Of course, theoretically it is possible, because in reality a purely capitalist society does not exist. The hierarchy here is objective, too. Marx, considering two-class capitalism in the first volume of *Das Kapital*, grasps the essence of capitalism.

Of course, the distinction of primary and secondary factors is not always so easy as in both examples cited above. The claim that a given factor is an essential, primary factor, is usually a hypothesis. Scientists build an ideal model of a considered process or object by means of this hypothesis. In other words, the construction of a model assumes a view about the essence of this process or object.

L. Nowak develops a whole theory of 'essentialist hierarchy' of magnitudes and of 'essentialist procedures' in science (Nowak, 1974a). This theory is complicated and some of its points seem to be not quite clear, but we shall not discuss it here.[9] We have presented the main idea which seems to be quite clear, although it is perhaps controversial too. We shall now present some other, more fundamental issues which also give birth to some controversies.

2.7. SOME CONTROVERSIAL ISSUES

L. Nowak claims in many papers that all genuine laws are idealizational — in any case in advanced sciences; the sole factual laws in them are the final concretizations of idealizational laws. I agree that all advanced sciences have to do with ideal models. We may even say that a branch of science becomes mature only when it uses ideal models and formulates idealizational laws. Nevertheless, it seems that Nowak's thesis is exaggerated.

First, there are *empirical laws* which are a direct generalization of experience, e.g. functional dependencies among some directly measured parameters, often presented in the form of a diagram. They are approximate — every measurement is made with an approximation. However, they are factual, for they do not assume any ideal models. We cannot expel these laws from science although we do not value them as highly as theoretical (usually — idealizational) ones.

Secondly, it is not clear whether all theoretical laws are idealizational. Different interpretations are sometimes possible. Consider the second Newton law of dynamics. The physicists say often that it is valid — like the first and third ones — in an inertial system. Indeed, in an inertial system the body having mass m on which force F acts gets the acceleration $a = (F/m)$. In the reality many forces (perhaps an infinity of forces) act on each body and when we regard the acceleration a caused by a given force F we must disregard all other forces, i.e. make an idealization. However, we may interpret this law otherwise. We may consider the resultant of all forces acting on a body and designate it by F. Then we obtain the empirical acceleration a and the law turns out to be factual. True, F ceases to be an empirical magnitude.

Another controversial issue is the problem of applying and testing idealizational laws. Nowak claims that an idealizational law cannot be applied to reality directly; the procedure of concretization is always necessary and only its final product may be applied and tested as being a factual law. According to him, overlooking this circumstance is a basic fault of all authors (inductivists, deductivists, Marxists) who recommend the empirical testing of laws without mentioning the necessity of their previous concretization (Nowak, 1971). I agree that the way described by Nowak is the main way of provability in advanced sciences and that the fault of almost all methodologists is the overlooking of it (see Ch. 6). Nevertheless, in many cases we may apply and test directly idealizational laws, namely, in cases in which real values of the characteristic parameters are small ($p_i \approx 0$) and an idealizational law describes reality with good approximation. Probably for each idealizational law there is a range of conditions in which $p_i \approx 0$ and the law may be applied directly with a sufficiently good approximation. Otherwise, Boyle and Mariotte or Ohm would not be able to test their laws which they held to be factual. For this reason the criticism of all previous methodologists made by Nowak, although in great measure legitimate, is exaggerated, which was pointed out in some reviews (e.g. Śleczka, 1972; Krajewski, 1972b).

Some authors who also highly appreciate Nowak's conception notice that he has not sufficiently explained the sources of knowledge about essential factors (Cackowski, 1975).

NOTES TO CHAPTER 2

[1] As is known, this law is called 'Boyle's law' in the Anglo-Saxon literature and 'lois de Mariotte' in French literature. If we regard the equation (the consequent) as the law, the first name is justified, if we take the whole implication, the second one is. The best solution is, of course, to call it 'Boyle-Mariotte law'; this name is used in Germany, Poland, Russia and other 'neutral' countries.

[2] Their idealizational character was only recently noticed in Marxist literature. G. Korch (1972, p. 320) writes that Marx has made numerous idealizing assumptions of different orders in *Das Kapital*. L. Nowak (1971, Ch. III) presents 27 idealizing assumptions made in the first volume of *Das Kapital*.

[3] The term *concretization* is closer to the Marxian expression. However, the term *abstraction* was abandoned by Poznań philosophers as ambiguous (every generalization is an abstraction but not every one is an idealization). After all, the term *concretization* is also ambiguous (we shall use it below in another meaning). The term *factualization* is not so ambiguous and using it we obtain a 'terminological harmony': idealization leads to idealizational laws, factualization to factual laws.

[4] The Poznań methodological school was founded by Jerzy Giedymin and is headed by Jerzy Kmita. Its most active member at present is Leszek Nowak.

[5] Strictly speaking, all three dimensions must equal to zero.

[6] Of course, we speak about a limit *cum grano salis*. We cannot form an infinite sequence of real bodies. Probably, there are no bodies with arbitrarily low positive values of given parameters (e.g., volume). However, we conventionally assume that there are some, that we can form an infinite sequence of these bodies, that the sequence is convergent, etc.

[7] Sometimes, to be sure, the characteristic parameters take on other values, e.g., the coefficient of light absorption of the perfect black body = 1. Noticing this, Subbotin criticized Gorski who wrote that the extreme value is zero. Nowak also uses a more general assumption: $p_i(x) = d$. However, we may always change the parameter so that we obtain zero. E.g., take the coefficient of reflection instead of the coefficient of absorption. When a parameter takes on positive and negative values, we may consider its absolute value. Hence, we do not lose the generality when we assume that $p_i = 0$ in the ideal objects and $p_i \geqslant 0$ in real objects.

[8] In fact, always $p_i(x) > 0$. However, we write $p_i(x) \geqslant 0$ in order to obtain more general laws.

[9] Some Polish logicians sharply criticized Nowak's concept of essential magnitudes and their hierarchy (e.g. Kaluszyńska, 1975).

CHAPTER 3

REDUCTION

3.1. THE CONCEPT OF REDUCTION

The term *reduction* has different meanings in philosophy (cf. Wartofsky, 1968, p. 347). It is often used in epistemology: positivists try to reduce abstract concepts to concrete ones, theoretical terms to observational ones, the whole experience to sense-data, etc. However, such a reduction is impossible and epistemological reductionism has very few adherents now.

We shall consider another concept of reduction which is alive in science and important to philosophy (though also raising controversies): the reduction of theories.

We say that a theory T_1 has been reduced to a theory T_2 when we can infer T_1 from T_2 or — which is usually the case — from T_2 and some additional assumptions. We shall consider the different kinds of assumptions.

Such a reduction allows explaining T_1 by T_2. Hence, reduction is a special case of explanation, more precisely of deductive-nomological explanation in which the *explanandum* is the logical consequence of the *explanans* (cf. Hempel and Oppenheim, 1948). Reduction is deductive-nomological explanation of theories (or laws). The reduced theory T_1 is the explanandum, the reducing theory T_2 (with additional assumptions) the explanans. We may call also T_1 *reducendum* and T_2 *reducens* (cf. Majewski, 1974).

We have designated the reduced theory by T_1 although it is logically secondary and therefore is designated by many authors as T_2. We have done so for chronological reasons. In the development of science the reducendum usually appears earlier than the reducens. We explain an old theory by a new one because the latter is usually more general (explanans must be, of course, not less general than explanandum). True, in the past, when people tried to explain phenomena animistically they did not use general laws but 'explained' unknown phenomena by known (or seemingly known) ones. The situation was similar with the mechanistic explanation, e.g. the consideration of an organism as a machine. However, in genuine scientific explanation we have an inverse relationship: we explain the better known by the worse known (e.g. the rainbow by the laws of optics) because we usually know particular phenomena better than general laws (cf. Feigl, 1964). The positivistic attempts to explain abstractions by sense-data, i.e. the worse known by the known, are as naive as the animistic 'explanations' (and are themselves animistic in essence since the psychological phenomena are considered as the final explanans).

There are two main types of theory reduction: *homogeneous* reduction when T_1 and T_2 are expressed in the same language (although one of them may contain more predicates) and *heterogeneous* reduction when the languages are different and we have to translate one into another (cf. Nagel, 1961, Ch. 11).

The attention of philosophers of science is usually concentrated on heterogeneous reduction. It is very important and raises many methodological questions. Nevertheless, we cannot agree with the opinion expressed by E. Nagel (*ibidem*) that homogeneous reduction is always trivial, uninteresting, and does not raise logical difficulties. We shall show logical difficulties in some kinds of homogeneous reduction below. But we shall consider heterogenous reduction first.

3.2. HETEROGENEOUS REDUCTION

Different logical problems connected with heterogeneous reduction are discussed by many authors (Kemeny and Oppenheim, 1956; Oppenheim and Putnam, 1958; Nagel, 1961; Hempel, 1965; Schaffner, 1967; Causey, 1972a, b; Zamiara, 1974). We shall not deal with them, limiting ourselves to the consideration of some general problems.

An especially important kind of heterogeneous reduction is reduction of a theory of a material system considered as a whole to a theory of its elements. It is often called *microreduction*. The postulate of microreduction has always been vivid in science and has often led to great successes (cf. Szumilewicz, 1970). A classic example, examined in detail by Nagel, is the reduction of thermodynamics (TD) to the molecular-kinetic theory of gases, finally to classical mechanics (CM). The portion of gas is a system of molecules moving according the laws of CM. Hence, TD is a theory of a system, CM a theory of its elements. Two sets of statements are necessary for the reduction of TD, besides CM. First, statements about the structure of the system: gas consists of a very large number of molecules which move in all directions with equal probability (so-called elementary chaos), collide as perfect elastic bodies, etc. We have statistical assumptions (elementary chaos) here; on their conjunction with CM the Classical Statistic Mechanics (CSM) is based; therefore, we may also say that TD is reduced to CSM. Second, there are necessary statements connecting the macroparameters of TD to the microparameters of CM (kinetic theory), namely equations expressing the temperature by the average kinetic energy of gas molecules, the pressure by the number of molecule impacts onto the unit of the surface of walls, etc.

In general, two sets of statements are necessary for the reduction:

(1) The description of the structure of the system (relations among its elements). We shall designate it by S.

(2) Statements connecting the terms of both theories, i.e. translatory rules. Sometimes they are interpreted as definitions of system concepts and are called 'coordinating definitions' (*Zuordnungsdefinitionen*, according to H. Reichenbach). More often, they are interpreted as synthetic sentences which connect terms defined independently. Often they are called 'rules of correspondence', according to Carnap, but this expression is not felicitous because of the ambiguity of the word 'correspondence' especially in the context of this book.[1] Therefore, I prefer another, also widespread, expression which will raise no confusion: *bridge rules*. We designate the conjunction of bridge rules by B. The scheme of microreduction is then, as follows:

(3.2.1) $T_2 \wedge S \wedge B \Rightarrow T_1$.

The rules B play a crucial role here. They allow translation of T_1 into T_2. There are different forms or kinds of bridge rules (cf. Causey, 1972a, b) but we shall not deal with this problem. In any case each discovery of B is not a result of semantic analysis but a result of scientific investigation (cf. Hempel, 1966, Ch. 8). It is possible not in an arbitrary chosen time but only when both theories are sufficiently developed (cf. Nagel, 1961, Ch. 11).

Some authors pose a question whether the rules B belong to T_1 or to T_2 (Brittan, 1970; Szmatka, 1975). Strictly speaking, they belong to none of them. They belong to a new theory T_2' which is a conjunction of T_2, S and B:

(3.2.2) $T_2' = T_2 \wedge S \wedge B.$

Hence,

(3.2.3) $T_2' \Rightarrow T_1$.

Nevertheless, we say that T_1 is reduced to T_2, according to the definition of reduction from 3.1.

3.3. NON-MECHANISTIC REDUCTIONISM

We may distinguish the internal (intra-disciplinary) and external (inter-disciplinary) reductions (cf. Kemeny and Oppenheim, 1956). The reduction of TD to CM is an example of the first type. However, the second type is more important from the philosophical point of view and raises many controversies.

One of the crucial questions in philosophy and science is the question whether it is possible to reduce the 'higher' sciences to the 'lower' ones, especially biology to physics and chemistry. Some scientists and philosophers maintain that such reduction is possible and desirable, others deny it. The former are traditionally called mechanists or reductionists, the latter vitalists, holists, emergentists. However, we shall not use the terms 'reductionism' and 'mechanism' as synonyms.

Classical mechanism, which reigned in science in the seventeenth, eighteenth and partly in the nineteenth centuries, tried to reduce all sciences to CM. In the second half of the nineteenth century it turned out that it is impossible to reduce even the whole of physics to CM and classical mechanism died out. The term 'mechanism' was widened then and used as a name of attempts (or postulates) to reduce biology to physics, psychology to biology, sociology to psychology – in general the 'higher' sciences to the 'lower' ones. The adversaries of such reduction always stressed the peculiarity of the higher systems and processes, saying that the whole is more than the sum of its parts, etc. These arguments were used not only by the vitalists and spiritualists but also by many materialistically oriented scientists and philosophers, including Marxists. F. Engels stressed the qualitative differences between the basic 'forms of movement' of matter and opposed the reduction of higher forms to lower ones; at the same time he pointed out the connection between different forms of movement, the weight of the lower sciences to the development of higher ones, the necessity of explanation of higher processes on the basis of lower ones, etc. His views were not always clear and sometimes allowed for different interpretation which raised many discussions among the Marxist philosophers (cf. Krajewski, 1973a, Ch. 4). For a long time an anti-reductionist view governed the scene, but in the last decades reductionism has been revived among Marxist philosophers of science (though not among Marxist philosophers in general). In Poland philosophers of science have spoken about the possibility of reduction for a long time (Kerszman, 1958; Eilstein, 1958; Amsterdamski, 1965; Majewski, 1967; Synowiecki, 1969; Szumilewicz, 1969). In the U.S.S.R. the leading philosophers of science are also saying at present that reduction in a certain sense is necessary; they criticize 'dogmatic' interpretations of the texts of the classics, sometimes they criticize some of Engels' expressions (e.g. that the lower processes only 'accompany' the higher as side-processes), etc. (Bazhenov, 1966; Zhdanov, 1968; Kedrov, 1969a; I.V. Kuznietsov, 1970a; Karpinskaya, 1971). This trend is convergent with the attitude of the majority of contemporary biologists, especially in the light of successes of genetics and molecular biology. One of the leading Soviet molecular biologists speaks about the 'triumphal march of reductionism in contemporary biology', though he admits the necessity of associating reductionist methods with 'integrationist' ones (Engelhardt, 1970). This trend is convergent with the attitude of the great majority of western philosophers of science.

The new reductionism, sometimes called 'reinterpreted mechanism' (Hempel, 1966), avoids the faults of the old mechanism. The new reductionism takes into account the peculiarity of the whole, it does not presuppose that laws of biology can be deduced from laws of physics and chemistry only. This is clear from our scheme (3.1.). The reduction of a biological theory T_1 to a physico-chemical theory T_2 is possible only when we use the description S of the biological system and the bridge rules B connecting biological terms with the physico-chemical ones. Both S and B contain biological terms. In other words, the specific biological terms must be contained not only in the reducendum but also in the reducens. In this sense we speak about non-mechanistic reduction. It demands an association of physico-chemical methods with biological ones.

Of course, the reduction of biological theories is only beginning in genetics and molecular biology and it is not yet possible at all in many other branches of biology (ecology, theory of evolution, ethology). We do not know whether reduction of all branches of biology will be possible in the future (in the way described above). We can only postulate attempts in this direction. If such a reduction succeeds, it is always a great achievement of science.

Of course, we do not postulate abandoning the peculiar biological methods, as some mechanists claim. All branches of biology must develop, both those in which reduction advances and those in which it is not yet possible. The mechanistic 'ban' of a peculiar biological method is harmful as is every 'police attitude'. More harmful, however, is the inverse 'police attitude': the warning that each reduction is 'mechanistic'. It was the attitude of Lyssenko in Soviet biology and of many philosophers. Luckily it now has been rejected.

We call 'mechanistic' not only a postulate of reduction to CM (mechanism in the narrower sense) but also a postulate of reduction to physics and chemistry (mechanism in the wider sense) when it disregards the necessity of S and B in the reduction. Our attitude − the postulate of reduction according the scheme 3.1 − may be called *non-mechanistic reductionism* (cf. Krajewski, 1974b).

3.4. TRIVIAL HOMOGENEOUS REDUCTION

Homogeneous reduction is sometimes trivial, indeed. Namely, in the case when T_2 is a simple enlargement of T_1. It is so in the case cited by Nagel: unification of the theories of the movement of celestial and terrestrial bodies into CM by Newton. In this case the old theory T_1 is a simple consequence of the new one T_2:

(3.4.1) $T_2 \Rightarrow T_1$.

To be more precise, we must add even here a simple assumption, namely the assumption that the extension of T_1 is a subset of the extension of T_2. We shall designate this 'extension statement' (cf. S. Nowak, 1970, p. 408) by E. The scheme takes the form:

(3.4.2) $T_2 \wedge E \Rightarrow T_1$.

Sometimes mentioning E is superfluous, e.g. when we reduce the law[2] that *celestial* bodies pull each other with a force proportional to the product of their masses and inversely proportional to the square of the distance between their mass-centers to the Newton law that all bodies pull each other in this way. Here E is an analytical sentence (each celestial body is a body). However, the situation is not always so simple. When we reduce the law 'All whales breathe with lungs' to the law 'Each mammal breathes with lungs', then E (each whale is mammal) is a synthetical statement which is not obvious — some people think that the whale is a fish.[3]

In any case we have here to do with a simple and trivial kind of reduction which raise no serious logical problems. However, there are other kinds of homogeneous reduction.

3.5. NON-TRIVIAL HOMOGENEOUS REDUCTION

Reduction in physics often concerns cases in which T_1 has not only a narrower range than T_2 but concerns a system of bodies with special relations called initial conditions (relations among these bodies at the beginning of considered process) and boundary conditions (relations between the system and its environment.)

A classic example cited in all manuals of physics is the reduction of three Kepler laws (KL) to the Newtonian CM (three laws of dynamics and the law of general gravitation). This example — and the example of the reduction of Galileo's free fall law to CM — is often discussed by philosophers of science. As we have said, already Duhem noticed logical problems here: KL are, strictly speaking, incompatible with CM and we can deduce from the latter only an approximation of the former. This remark is repeated by many contemporary authors (Hempel, 1965; Kemeny, 1959; Popper, 1957). However, a question arises: may we speak here of reduction? Some philosophers answer: no! Feyerabend blames Nagel and other representatives of the 'orthodox' philosophy of science for neglecting the incompatibility of Galileo's and Kepler's laws with CM (Feyerabend, 1962a). Indeed, the analysis contained in Nagel's book is incomplete. Nagel admitted it later himself but he said only, as many others, that the approximation of Galileo's and Kepler's laws may be inferred from CM (Nagel, 1970). W.D. Siemens noticed that we must consider two relations here: a relation of correction of an earlier theory and a relation of reduction of the corrected theory to a later (more general) one (Siemens, 1971). This idea is right, but in order to carry it out we must consider the additional conditions of reduction and take into account the idealizational nature of KL in the light of CM. Only then can we get free from approximation.

We shall examine only reduction of the first KL stating that a planet revolves along an ellipse around the Sun being in one of the ellipse focuses. In order to obtain this law from CM we must assume the following initial and boundary conditions (rarely mentioned explicitly by physicists).

(1) Either the system consists of the Sun and *one* planet, or the other planets do not act (otherwise, the orbit would not be exactly elliptic because of perturbations caused by their gravity).

(2) No external forces act on the system (they would disturb the orbit, of course, too).

(3) The mass of the Sun is infinitely bigger than the mass of the planet (only then the center of the Sun is exactly in the focus of the ellipse).

(4) The distance of the planet from the Sun exceeds some minimal value — the radius of Roche's zone (otherwise the planet would be disrupted by the Sun).

(5) The tangent component of the initial velocity of the planet exceeds a minimum (otherwise, the planet would fall into the Sun).

(6) It does not reach a maximum (otherwise, the planet would move along a hyperbola and leave the Sun system).

We may briefly write these conditions as follows:

(3.5.1) $F_{pl} = 0$

(3.5.2) $F_{ext} = 0$

(3.5.3) $\dfrac{m_p}{m_s} = 0$

(3.5.4) $R(p,S) > R_{min}$

(3.5.5.–3.5.6) $v_{min} < v_{tang} < v_{max}$.

The first three assumptions (equalities) are idealizing ones, the last three (inequalities) are factual ones. Hence, the first Kepler law is, in the light of CM, idealizational. So also are other KL. They are fulfilled in real planet orbits only approximately.

The six cited assumptions describe the initial and boundary conditions or, in other words, the structure of considered ideal system. We shall designate their conjunction by S. Then the scheme of the reduction is as follows:

(3.5.7) $T_2 \wedge S \Rightarrow T_1$.

Notice that the vocabulary of T_2 may be richer than the vocabulary of T_1: In our example there are different terms (dynamic ones) in T_2 (CM) which are not used in T_1. On the other side, in this case there are in T_1, apart from purely mechanical terms (distance, velocity, etc.), such additional astronomical terms as 'Sun' and 'planet'. However they are not qualitatively different from the terms of T_2 and can be easily replaced by purely mechanical terms such as 'free body with a big mass', etc. A translation is not needed — the reduction is homogeneous.

After all, the contemporary form of KL is more general. As is known, these laws describe the revolutions of all celestial bodies — from sputniks to double stars. In the latter case the masses of bodies often do not differ much. Therefore, according to the first KL in a general form, the mass center of a system of two celestial bodies is in a focus of the ellipses along which both bodies move (the ellipses are, of course, different, they have only a common focus). Then the condition (3.5.3) is superfluous. In this form the particular terms 'Sun' and 'planet' disappear.[4]

We can see now that we have got rid of approximations. KL are idealizational laws, hence they are not false but emptily fulfilled in reality. The factualizations of KL are non-emptily fulfilled in real systems. In this interpretation KL are not incompatible with CM. On the contrary, KL as idealizational laws follow from CM and the additional assumptions (3.5.1)–(3.5.6).

We do not notice here the logical difficulties mentioned in 1.5 with respect to QM and STR. The idealizing assumptions (3.5.1) and (3.5.2) do not violate any assumptions of CM. Moreover, analogous idealizing assumptions are made in CM itself. The Newton law of general gravitation (in its elementary form) considers the gravitational force between two isolated bodies; it neglects the existence of other bodies and other (non-gravitational) forces although they always exist in reality.

The law of general gravitation and KL have the same 'degree of idealization'. Only their domains are different: the domain of the former is the set of all pairs of material bodies, the domain of the latter the set of pairs of free celestial bodies, i.e. the former is more general than the latter. We assume the following definition: a law L_2 is more general than a law L_1 if and only if the domain D_1 of L_1 is a subset of the domain D_2 of L_2:

(3.5.8) $D_1 \subset D_2$.

The comparison requires the same degree of idealization of both laws.

3.6. REDUCTION OF AN IDEALIZATIONAL LAW TO A FACTUAL ONE

A special kind of non-trivial homogeneous reduction is the reduction of an idealizational law to a factual one.

As we know from 2.5, the equation $F(x) = 0$ of an idealizational law is usually simple and the equation $F'(x) = 0$ of a factual (or more factual) law is more complicated. When we introduce the idealizing condition C_I ($p_i = 0$) we pass back from L_F to L_I:

(3.6.1) $L_F \wedge C_I \Rightarrow L_I$.

When we consider only equations, we may write:

(3.6.2) $F'(x) = 0 \wedge p_i(x) = 0 \Rightarrow F(x) = 0$.

The reduction is, of course, homogeneous because L_I and L_F are expressed in the same language.

The reduction of L_I to L_F is obviously possible. It is not clear whether the opposite reduction of L_F to L_I is also possible. L. Nowak claims that it is, that L_F follows from L_I and some additional knowledge. Undoubtedly scientists in their theoretical reasoning usually start from L_I and pass to L_F, taking into account factors disregarded earlier. But is it a deduction? Nowak has not analyzed the nature of the additional knowledge and has not shown that the deduction of L_F from L_I is indeed always possible (cf. Ślęczka, 1972).

Nevertheless, we cannot exclude the possibility of the reduction of L_F to L_I. We see, by the way, that the concepts of reduction and explanation of theories should not be identified. Maybe Nowak is right and reduction is possible in both directions (of course, with different additional assumptions). Shall we then say that L_I explains L_F and L_F explains L_I? It would be a very strange conclusion.

In practice we do not say that L_F explains L_I, e.g. that the van der Waals law explains the Boyle-Mariotte one or that the differential law of current explains the Ohm one. Rather the opposite is true: in order to better understand a more complicated law L_F we start from a simpler law L_I and gradually add additional factors. We explain L_F by L_I though we are not sure whether there is a strict reduction here. We do not say that L_I is explained by L_F though there is an obvious strict reduction here.

We do explain L_I but not by means of L_F. We explain a L_I by reducing it to a more general L_I. E.g. we explain the Boyle-Mariotte law by reducing it to CM but not to the van der Waals law.

In any case reduction and explanation are not identical. This justifies the introduction of the terms *reducendum* and *reducens*.

There are in many cases serious logical difficulties in the reduction of L_I to L_F. We shall consider them in the next chapter, which treats the correspondence relation which is also a relation between L_F and L_I.

NOTES TO CHAPTER 3

[1] Some authors notice that there is a connection between two meanings of this term: Bohr's CP connects theoretical terms of the new quantum theory to the observational terms of the classical theory of radiation (cf. Such, 1974).

[2] A theory is usually a system of laws, definitions, etc. We may consider, however, a single law as a special case of the theory.

[3] In Polish the name for whale 'wieloryb' means 'big-fish'.

[4] The KL often cause trouble for philosophers of science because they contain a proper name, 'Sun'. In the contemporary form of KL this trouble disappears.

CHAPTER 4

CORRESPONDENCE RELATION

4.1. DEFINITION

We shall now examine the concept of correspondence relation (CR) more thoroughly.

As we know from 1.3, according to many Soviet authors CR takes place not only in physics but also in mathematics. I agree that it is possible to give a comprehensive definition of this relation. However, in order to diminish the number of difficulties I prefer to limit its extension to physics or, strictly speaking, to the empirical sciences. We draw examples from physics but probably analogous relations take place in other empirical sciences when they are at a sufficiently advanced stage of development (perhaps this is a problem of the future).

CR is a relation between two theories T_1 and T_2 (or two laws L_1 and L_2) of an empirical science. We shall call T_1 (or L_1) *corresponded* and T_2 (or L_2) *corresponding*.

The intension of the concept of CR contains two parts: a temporal and a logical one. The first is simple. The CR assumes a temporal succession: T_1 is formulated at a time t_1 and T_2 at a later time t_2. This requirement is necessary if we want to give the CR a normative value. We may write:

(4.1.1) $t_1 < t_2$.

The logical part is more complicated and entails numerous controversies. The CR is connected to the relation of reduction. Should we identify its logical part with this relation? Some authors assume that the CR takes place between each pair of laws L_1 and L_2 when L_1 is formulated earlier than L_2 and when we can reduce L_1 to L_2 (or, in any case, the equation of L_1 to the equation of L_2). They consider, e.g. the relation of KL to CM as a CR (I. Nowak, 1972, Szumilewicz, 1974; Such, 1974). I think this approach is too comprehensive. It disregards a feature usually held essential for the CR: the asymptotical transition from the corresponding law to the corresponded one. In the case of the relation of KL to CM there is, of course, no such transition.

We shall now formulate the logical part of the definition of CR for laws (logical condition).

A law L_1 is in a CR with a law L_2 if the equation $F_1(x) = 0$ of L_2 passes asymptotically into the equation $F_1(x) = 0$ of L_1 when some characteristic for L_2 parameters p_i tend to zero ($p_i \rightarrow 0$). If we assume that they reach the limit ($p_i = 0$), we may deduce the equation of L_1 from the equation of L_2:

(4.1.2) $F_2(x) = 0 \wedge p_i(x) = 0 \Rightarrow F_1(x) = 0.$

We can now sum up the definition of CR. A law L_2 is in a CR with a law L_1 if and only if conditions (4.1.1) and (4.1.2) are fulfilled.

The scheme (4.1.2) is identical with the scheme (3.6.2). Indeed, the corresponded law is always idealizational and the corresponding law is more factual. We can see it in all considered examples.

To be more precise, the corresponded law L_1 is usually held by its discoverer to be a factual one but it turns out to be idealizational in the light of L_2. We designate by L_1' the law L_1 reinterpreted in the light of L_2. Then L_1' is always an idealizational law and L_2 a more factual one.

We have said that the equation of L_1 can be reduced to the equation of L_2. The problem whether such a reduction is possible for whole laws L_1 and L_2 will be discussed below in this chapter.

We spoke about laws. In the case of theories the formulation of the logical condition must be more careful. We may say that the CR between T_1 and T_2 takes place if it takes place between some of their basic laws. We must not speak about *all* laws. As we know, a classic example of CR is the relation of STR to CM. Many equations of STR pass into equations of CM when $1/c \rightarrow 0$. However, not all: the equation $E = mc^2$ does not.

4.2. SIMPLE IMPLICATIVE VERSION

We pass to the relation among whole laws. The above-mentioned logical difficulties arise here. In order to avoid them, different versions (interpretations) of the CR have been proposed. We shall make a survey of them beginning with the version which does not notice these difficulties – a simple implicative version.

Some physicists and even philosophers say that in the case of CR the old law L_1 is a special case of the new law L_2. In other words, they hold

L_1 to be a simple consequence of L_2:

(4.2.1) $L_2 \Rightarrow L_1$.

Others (e.g. Kuznietsov, 1948) stress the weight of the limit condition. According to them, L_1 is a consequence of L_2 provided $p_i = 0$:

(4.2.2) $L_2 \wedge p_i = 0 \Rightarrow L_1$.

This scheme is not so naive as the scheme (4.2.1). It is correct in a syntactical respect, i.e. when we take into account only mathematical equations. However, when we consider laws as implications we must reject such a simplified solution. As we know, the condition $p_i = 0$ is always false. Hence the whole antecedent of (4.2.2) is false. Sometimes, it is simply empirically false, sometimes, apart from this, it contradicts the basic principle of T_2 (supposed also by L_2). L_1 and L_2 are incompatible and the reduction, in such a simple way, is impossible.

Therefore, the simple implicative version of CR fails.[1] We may add that it supposes a simple cumulative conception of the development of science. It disregards the necessity reinterpreting an old theory in the light of a new one.

Now we pass to those versions of CR which take into account the contradiction between L_1 and L_2.

4.3. APPROXIMATIVE VERSION

Many authors use Duhem's idea and say that L_1 is incompatible with L_2, nevertheless an approximation of the former follows from the latter.

If we designate the approximation of L_1 by aL_1 we may present the approximative version in the following two implications:

$$(4.3.1) \begin{cases} L_2 \Rightarrow \neg L_1 \\ L_2 \Rightarrow aL_1. \end{cases}$$

These conditions are admitted also by some authors who do not talk about correspondence but about approximate reduction or approximate explanation (e.g., Tuomela, 1973, Schiebe, 1973). However, they usually do not explicate more closely the concept of approximation. Sometimes they say that aL_1 is similar or analogous to L_1 (e.g. Schaffner, 1967) but this does not explain much. Some authors suggest use of the method of successive approximations but do not develop this idea (Augustynek, 1974).

It is obvious that every approximation has a limited value. Many authors point out that an approximation is more or less satisfactory in some limited area (in which the old theory had been proved) but not in a broader area revealed by the new theory (e.g., Groenwald, 1965). Indeed, L_1 gives a good approximation to reality when the parameters p_i characteristic for L_2 may be neglected: h in the macroworld, $1/c$ when the velocity is low ($v \ll c$), etc.

In order to formulate the approximative version more precisely we may introduce a concept of approximative implication or approximative consequence which is used as a matter of fact in mathematics and physics in approximative calculations (cf. Zytkow, 1974). Consider a basic magnitude a contained in L_1 and L_2. Suppose that the precision of our measurement (or calculation) allows for the values of a errors not exceeding ϵ. Suppose later that the equation $F_1(x) = 0$ of L_1 follows from the conjunction of the equation $F_2(x) = 0$ of L_2 and the assumption $p(x) = 0$. Designate by $a_1(x)$ and $a_2(x)$ the values of a of a body x predicted respectively by L_1 and L_2. Then we shall say that L_1 follows approximately from L_2 when the absolute value of the difference $a_2(x) - a_1(x)$ is always smaller than ϵ:

(4.3.2) $\bigwedge_x |a_2(x) - a_1(x)| < \epsilon$.

This condition is fulfilled if p is sufficiently low, lower than some δ:

(4.3.3) $p(x) < \delta$.

The value of δ is correlated with the value of ϵ. The smaller ϵ is, i.e. the more precise our measurement and calculation is, the lower δ is, i.e. the smaller range of phenomena that should be considered when we want to obtain the approximative implication.

We may identify the approximative inference of L_1 from L_2 with the inference of an approximation of L_1 from L_2.

The whole domain D_2 of L_2 should be divided into two parts: D_1 in which L_1 gives a satisfactory approximation ((4.3.2) and (4.3.3) are here fulfilled) and $D_2 - D_1$ in which it does not ((4.3.2) and (4.3.3) are not fulfilled). The division depends, of course, on the value of ϵ. The smaller ϵ is (and δ), the smaller D_1 is.

Strictly speaking, a negation of L_1 follows from L_2 in the whole D_2. An approximation of L_1 follows from L_2 in D_1. A negation of this approximation follows from L_2 in $D_2 - D_1$. We may write:

$$(4.3.4) \quad \begin{cases} L_2 \Rightarrow \neg L_1 & \text{in whole } D_2 \\ L_2 \Rightarrow aL_1 & \text{in } D_1 \\ L_2 \Rightarrow \neg aL_1 & \text{in } D_2 - D_1 \end{cases}$$

The approximative version of CR is much better than the simple implicative one and is useful for many aims (we shall use it in Ch. 5). It breaks with simple cumulativism because it takes into account the fact that we must give a new interpretation of L_1 in the light of L_2: now L_1 is interpreted not as an exact law (as it was interpreted before L_2 was discovered) but as an approximate one. Its domain is narrower.

However, the approximative version disregards the necessity of a more basic reinterpretation of L_1 in the light of L_2: L_1 turns out to be an idealizational law. Therefore, this version is not quite satisfactory.

4.4. EXPLANATIVE VERSION

Many authors abandon the implication at all and choose another way to confront L_1 and L_2: the comparison of their explanative powers. Therefore, we call this version an *explanative* one.

There are two variants of the explanative version of the CR. One of them may be called a *qualitative* one, another a *quantitative* one.

A. *A Qualitative Variant*

According to the first variant, L_2 is in a CR with L_1 if L_2 explains all phenomena explained by L_1 and some additional phenomena which were not explained by L_1. If we designate by $\text{Expl}(L_1)$ the set of phenomena explained by L_1 and by $\text{Expl}(L_2)$ the set of phenomena explained by L_2, we may formulate this condition as follows:

$$(4.4.1) \quad \text{Expl}(L_1) \subset \text{Expl}(L_2).$$

Some authors formulate this condition as a definition of the reduction of one theory to another — without historical aspects (e.g. Kemeny and Oppenheim, 1956).

Other authors analyzing the regularity of the history of science accept this condition as a necessary one but do not call it 'correspondence' (e.g. Lakatos, 1970).

Stefan Amsterdamski treats this condition as the condition of *cumulation* of knowledge and not of correspondence, which he interprets implicatively; therefore, he claims that there is a cumulation but there is no correspondence in the course of the development of science (Amsterdamski, 1973).

Some authors connect the explanative version of CR with the implicative one; they say that the new theory 'not only explains new phenomena but contains the old one as a limit case' (Misiek, 1969). We may say that they admit the implicative version of CR (the explanative version follows from the implicative one).

B. A Quantitative Variant

The quantitative variant of the explanative version of CR takes into account the precision of explanation (or prediction) given by the corresponded and the corresponding laws (theories).

According to it, L_2 is in a CR with L_1 if L_2 explains (predicts) phenomena in a domain D better (and in other domains not worse) than L_1.

When we want to formulate it more precisely, we must introduce some metrical concepts. We shall present here a general idea of this precisioning.

We designate again by $a_1(x)$ and $a_2(x)$ the values of a of the object x predicted respectively by L_1 and L_2. We introduce now the concept of *real* value $a_0(x)$; we may interpret it empirically as an average value in a set of measurements. We shall say that L_1 gives the prediction of a in D with the precision ϵ_1 when the absolute value of deviation of a_1 from a_0 in D is always smaller than ϵ_1:

$$(4.4.2) \bigwedge_{x \in D} |a_1(x) - a_0(x)| < \epsilon_1.$$

Analogously for L_2:

$$(4.4.3) \quad \bigwedge_{x \in D} |a_2(x) - a_0(x)| < \epsilon_2.$$

We say that L_2 is in a CR with L_1 when

$$(4.4.4) \quad \epsilon_2 < \epsilon_1.$$

A more precise formalization of this version may be achieved in different ways. Some Polish logicians do it by means of model-theoretical methods elaborated for analysis of the language of empirical sciences by R. Wójcicki and M. Przełęcki. They use especially Wójcicki's concept of operational structure and his approximative concept of truth (see Wójcicki, 1974) and treat CR as a relation among theories whose operational structures give different degrees of approximation to a fragment of reality; they also use different additional concepts (Garstka, 1974; Nadel-Turoński, 1974). We see that some operational concepts are used here, which is probably inevitable in this case.

I evaluate very highly the quantitative explanative version of CR. It expresses in a precise way some important intuitions of physicists, but nevertheless, not all of them. It neglects the internal logical relations among laws (and theories) which appear in the course of the development of science. The following versions try to take them into account.

4.5. 'DIALECTICAL' VERSION

Izabella Nowak in her doctoral thesis criticizes all earlier interpretations of CR for neglecting the idealizational nature of scientific laws and the logical problems arising from it. She develops a new interpretation, calling it a *dialectical* one because it is inspired by Marxian ideas. We cite her definition (after I. Nowak, 1974) making some minor simplifications.

The idealizational law T' dialectically corresponds to the (idealizational or factual) law T if and only if:

(1) T has been formulated earlier than T'.

(2) T has the same consequent as T'.

(3) T has the same factual assumptions of T'.

(4) The set of idealizing assumptions of T is a subset of the set of idealizing assumptions of T'.

(5) It is possible to obtain a concretization (in our terminology 'factualization') T'' of T, which has the same idealizing assumptions as T.

The symbols used here correspond to ours: T to our L_1, T'' to L_2, T' to L_1'. We see that for I. Nowak the CR takes place not between L_1 and L_2 but between L_1 and L_1'. We shall illustrate this with a simple example. Take the law of addition of two velocities having the same direction. In CM its equation is:

$$v = v_1 + v_2.$$

In STR:

$$v = \frac{v_1 + v_2}{1 + \dfrac{v_1 v_2}{c^2}}$$

The STR accepts the (factual) assumption that $c = $ const. At the same time it reveals that the equation of CM is valid only under the idealizing assumption $c = \infty$. We shall present three schemes of laws omitting the assumptions common to all three laws and writing only those assumptions which distinguish these laws from each other.

L_1 (T): $\qquad\qquad\qquad v = v_1 + v_2$

L_1' (T'): $\quad c = \infty \quad \Rightarrow \quad v = v_1 + v_2$

L_2 (T''): $\quad c = $ const. $\Rightarrow \quad v = \dfrac{v_1 + v_2}{1 + \dfrac{v_1 v_2}{c^2}}$

I. Nowak says that she considers the relation between T and T' (and not T and T'') as the CR because she holds the idealizational law T' to be more 'basic' than the factual law T''. But this procedure is in sharp contradiction to the language customs of physicists (which she admits). They would say that T and T' present the same law. In Mejbaum's terminology we have here two expressions of one law (see 2.1). Therefore, it is better to consider CR to be the relation between L_1 and L_2, i.e. between an earlier law which turned out to be an idealizational one and its later factualization. These laws have different consequents, hence they are by all authors considered as *different* laws.

In order to pass from a corresponded law L_1 to the corresponding one L_2 we must take two steps: (1) to reveal the idealizing assumptions (the transition from L_1 to L_1'), (2) to abrogate these assumptions and obtain again a factual[2] law L_2 with a more complicated equation (see Such, 1974; Krajewski, 1974).

I. Nowak shows rightly the faults of the simple implicative version (and some other versions) of CR. Her doctoral thesis inspired some Polish philosophers, including me, to analyze CR more closely. Nevertheless, I could not agree that CR in a proper form has nothing to do with implication, as I. Nowak claims. I wanted to 'save' the intuitions of physicists who say that L_1 'follows' from L_2. For this reason I have proposed a new interpretation of CR (Krajewski, 1973c, 1974a). It may be called a *renewed implicative* one.

4.6. RENEWED IMPLICATIVE VERSION

We shall examine further the law of addition of two velocities with the same direction. As we know, L_2 supposes the assumption $c = $ const. — the basic assumption of the whole STR. On the contrary, $L_1{}'$ presupposes the assumption $c = \infty$ which is incompatible with the former. Therefore, we cannot deduce the law of CM (in its contemporary interpretation, i.e. $L_1{}'$) from the law of STR (L_2). Nevertheless, this procedure will be possible when we perform an additional operation on L_2. I call this procedure abstracticization. We abrogate the assumption $c = $ const. of L_2 without introducing the assumption $c = \infty$ of L_1. We obtain then a law which has neither of these assumptions and designate it by L_2^*. Hence, there are the following four versions of the velocity-addition law (we still omit common assumptions):

L_1	CM		$v = v_1 + v_2$
$L_1{}'$	CM in the light of STR	$c = \infty$	$\Rightarrow v = v_1 + v_2$
L_2	STR	$c = $ const.	$\Rightarrow v = \dfrac{v_1 + v_2}{1 + \dfrac{v_1 v_2}{c^2}}$
L_2^*	Abstracticized STR		$v = \dfrac{v_1 + v_2}{1 + \dfrac{v_1 v_2}{c^2}}$

How should we interpret L_2^*? It is already not a law of STR (we have abrogated its basic assumption) and not yet a law of CM (we do not introduce its idealizing assumption). It is an equation but, of course, not a simple mathematical equation: its terms are physically interpreted (v_1, v_2, v, c are velocities). However, this interpretation is not complete. We have an 'intermediary product' of reasoning.

To the 'abstracticized' law L_2^* we may add the idealizing condition $c = \infty$ without contradiction and thus obtain L_1':

$$L_2^* \wedge c = \infty \Rightarrow L_1'.$$

Generalizing this whole procedure, we obtain the following table:

L_1	Corresponded law	$F_1(x) = 0$
L_1'	Corresponded law reinterpreted in the light of corresponding one	$p_i = 0 \Rightarrow F_1(x) = 0$
L_2	Corresponding law	$p_i > 0 \Rightarrow F_2(x) = 0$
L_2^*	'Abstracticized' corresponding law	$F_2(x) = 0$

The scheme of reduction is then as follows:

(4.6.1) $L_2^* \wedge p_i = 0 \Rightarrow L_1'$.

Notice that L_1 is formulated in an earlier period of the development of science (I), and L_2 and L_1' in a later period (II).

Consider now the relations among the four members of the table presented above. It will be more convenient to consider the transformations of one into another. The transformation of L_1 into L_2 is *correspondence* – it adds to the antecedent a factual assumption and changes the consequent $F_1(x)$ into $F_2(x)$. The transformation of L_2 into L_1' is *idealization* – it changes in the antecedent some factual assumptions into idealizing ones and the consequent $F_2(x)$ into $F_1(x)$. The inverse transformation is *factualization* – it changes some idealizing assumptions in the antecedent into factual ones and $F_1(x)$ into $F_2(x)$ in the consequent. The transformation of L_2 into L_2^* is *abstracticization* – it eliminates the factual assumption from the antecedent without any change of the consequent. The inverse transformation is factualizing *concretization* – it reintroduces the factual assumption into the antecedent without any change of the consequent. The transformation of L_2^* into L_1' may be called *idealizing concretization* – it introduces the idealizing assumption into the antecedent and changes $F_2(x)$ into $F_1(x)$ in the consequent. At last, the transformation of L_1 into L_1' is a *revealing of idealizing assumptions* of the equation $F_1(x) = 0$ (which is realised only in the period II). All these relations are shown in the graph on p.51.

There is a semantical problem that we have not yet considered. Have symbols in L_2 the same meaning as in L_1? We shall discuss this problem in Ch. 5.

4.7. SOME FORMAL FEATURES

A. *Irreflexivity*

CR is *irreflexive*. It follows directly from the temporal condition (4.1.1)

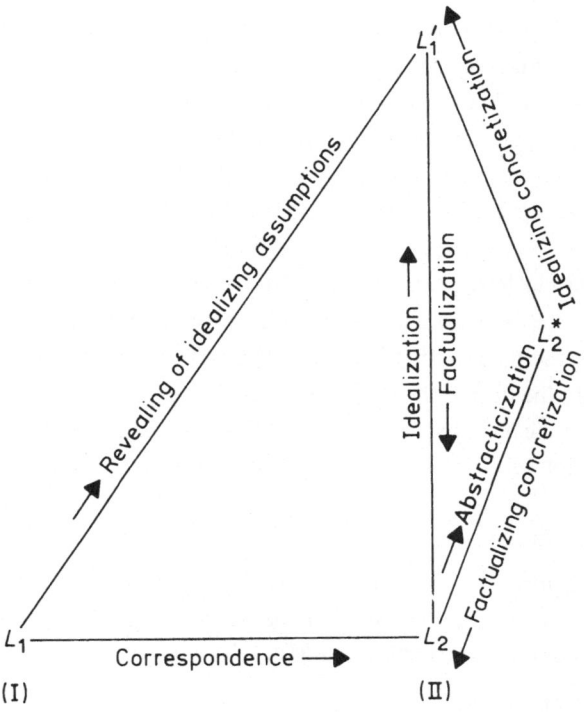

which contains a sharp inequality.

Of course, if we used a non sharp inequality ($\leqslant$), CR would be reflexive. There are authors who choose this way (Garstka, 1974). However, it seems to be unnatural: nobody says that a theory corresponds to itself.

B. *Asymmetry*

CR is asymmetric. It follows both from the temporal condition (4.1.1) and the logical condition (4.1.2). The implication is not possible in the opposite direction: we cannot deduce L_2 from L_1 and a limit condition.

The asymmetry between L_1 and L_2 is connected to different degrees of their generality. Usually one says that the corresponding theory (or law) is more general that the corresponded one. Some authors show that this is not true (in a literal sense). Since L_1' and L_2 have different and incompatible conditions (antecedents), none of them is more general than another, and they cannot be compared in this respect (I. Nowak, 1974). However, we need not stop at this negative conclusion. The equation $F_2(x)$ is more general than the equation $F_1(x)$. The abstracticized law L_2^* is also more general than L_1' (the latter contains a limiting condition). In this not literal sense we may say that L_2 is more general than L_1'. Strictly speaking, we ought to say that L_2^* is more general than L_1'.

We meet an interesting paradox here. L_2 is more factual than L_1' and, at the same time, more general (if we 'abstracticize' it). The factualization (the concretization, according to the terminology of the Poznań school) is associated with the generalization! This paradox is connected to another one: an idealizational law says more about reality than an empirical factual law. Hence, we have a double paradox here (cf. Ślęczka, 1972).

The asymmetry of CR has an important methodological value. We may evaluate the deviation of L_1 from experience and the scope of its sufficiently good application by means of L_2. However, we cannot do the opposite: an evaluation of L_2 by means of L_1 is impossible (cf. Szumilewicz, 1974).

C. *Transitiveness*

CR is *transitive*: when there is a CR between L_1 and L_2 and between L_2 and L_3, then there is also a CR between L_1 and L_3. The transitiveness of the temporal condition is obvious:

(4.7.1) $t_1 < t_2 \wedge t_2 < t_3 \Rightarrow t_1 < t_3$.

The logical condition is also transitive. Consider successive laws L_1, L_2, L_3. Their equations have shapes: $F_1(x) = 0$, $F_2(x) = 0$, $F_3(x) = 0$. We designate L_2 reinterpreted in the light of L_3 by L_2' and L_1 reinterpreted in the light of L_3 by L_1''. As we know, L_2 reveals the idealizing assumption $p_i(x) = 0$ of L_1'. Further, L_3 reveals the idealizing assumption $p_j(x) = 0$ of L_2'. Hence, L_3 reveals the double idealizing assumptions of L_1'': $p_i \wedge p_j$. The laws L_1'', L_2', L_3 have decreasing degrees of idealization. The following implications take place:

(4.7.2) $F_3(x) = 0 \wedge p_j(x) = 0 \Rightarrow F_2 = 0$

(4.7.3) $F_2(x) = 0 \wedge p_i(x) = 0 \Rightarrow F_1 = 0$

(4.7.4) $F_3(x) = 0 \wedge p_i(x) = 0 \wedge p_j(x) = 0 \Rightarrow F_1(x) = 0$.

We can easily see that (4.7.4) follows from the conjunction of (4.7.2) and (4.7.3).

As an example we shall take CM, QM and Dirac's RQM. As we know, RQM passes asymptotically into QM when $1/c \to 0$ and QM passes asymptotically into CM when $h \to 0$. We may write the following short, though not precise, implications (to be more precise, we must take the abstracticized laws of these theories or only their equations):

$$RQM \wedge 1/c = 0 \Rightarrow QM$$

$$QM \wedge h = 0 \Rightarrow CM$$

Hence, $RQM \wedge 1/c = 0 \wedge h = 0 \Rightarrow CM$.

We may pass from RQM to CM also in another way, through STR:

$$RQM \wedge h = 0 \Rightarrow STR$$

$$STR \wedge 1/c = 0 \Rightarrow CM$$

Hence, $RQM \wedge h = 0 \wedge 1/c = 0 \Rightarrow CM$.

We see from this example that CR is not one-to-one correspondence and not even a simple unique correspondence. One theory may correspond to two theories: e.g. RQM to QM and STR. Two theories may correspond to one theory: e.g. STR and QM to CM.

4.8. CORRESPONDENCE SEQUENCE AND CORRESPONDENCE NETWORK

The transitiveness of CR enables the formation of *correspondence sequences* (cf. Niedzwiedzki, 1974; Such, 1974) of theories.

Consider a sequence of successive theories:

(4.8.1) $T_1, T_2, T_3 \ldots$

There is in such a sequence a CR between every pair of its members, both neighbor and (thanks to transitiveness) non neighbor. Always the antecedent member is a corresponded theory and the subsequent one — the corresponding theory.

When we consider the theories $T_1, T_2, T_3 \ldots$ in their original interpretation, they may be all factual (i.e. held to be factual by their creators) or have a low degree of idealization. When we consider, however, these theories in contemporary interpretation, i.e. as reinterpreted in the light of the last theory, the degree of their idealization gradually decreases. The last theory is held to be factual (probably until a new member of this sequence is formulated . . .).

We may form also another sequence — a sequence of reinterpretations of the corresponded theory in the light of successive corresponding ones:

(4.8.2) T_1, T_1', T_1'' ...

It is a sequence of successive revealing of idealizing assumptions of T_1. The degree of idealization gradually increases in this sequence.

We shall return to sequences (4.8.1) and (4.8.2) in Ch. 8.

Since CR is not a simple unique relation, both considered sequences may branch (diverge), later on converge again, etc. We obtain a *correspondence network* (cf. Niedzwiedzki, 1974). Such a network based on CM is presented in the following graph.

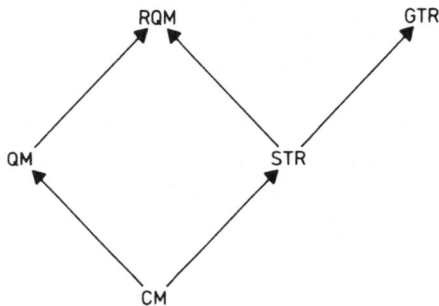

NOTES TO CHAPTER 4

[1] I came to this conclusion only in 1972. Till this time I held the scheme (4.2.2) to be correct.

[2] Strictly speaking, L_2 is more factual than L_1'. We may add that L_2 has the same degree of idealization as L_1.

CHAPTER 5

THE PROBLEM OF THE INCOMMENSURABILITY AND RELATIONS AMONG THEORIES

5.1. THE CLAIM OF INCOMMENSURABILITY

The whole search for logical relationships among theories has been threatened by some philosophers of science who claim that rival theories in science are usually (or always) *incommensurable*. T.S. Kuhn says that two theories divided by a scientific revolution are incommensurable because the meaning of the terms and even the whole world-image change radically in the course of the revolution (Kuhn, 1962). P.K. Feyerabend claims that all rival theories, even simultaneous ones, use different concepts and different language, and therefore are incommensurable (Feyerabend, 1962a). Later on Feyerabend noticed the temporal coincidence of the word 'incommensurable' as employed by him and Kuhn, at the same time pointing out certain differences between them (Feyerabend, 1970). Some other philosophers of science expressed similar ideas even earlier: N.R. Hanson who stressed the *theory-ladenness* of facts, the determination of each observation by a theory (Hanson, 1958), S. Toulmin who claimed that phenomena are determined by an ideal of natural order (Toulmin, 1961) and some others.

Many philosophers of science criticized this view. C.R. Kordig analyzed it in a separate book calling it RMV (Radical Meaning Variance view) (Kordig, 1971). We shall adopt this abbreviation.

The main argument used by Kordig and many other critics (Achinstein, 1964; Fine, 1967; Putnam, 1965, 1973; Koertge, 1969; Sellars, 1973, etc.) is the following one. RMV leads to the destruction of each logical analysis of science. It precludes the logical comparison of theories: if they are incommensurable, then they are neither compatible nor incompatible. The adherents of RMV, however, often speak about the incompatibility of rival theories, hence they contradict themselves. Sometimes they say, in order to avoid criticism, that two incommensurable theories, though not incompatible, may lead to incompatible experimental results (Feyerabend, 1965b). However, this claim presupposes the possibility of a separation of observational and theoretical terms which is always denied by RMV theorists, whereby they again contradict themselves (cf. Fine, 1967). The RMV precludes any revision of a theory by means of experimental evidence: if each observation is determined by a theory, it cannot be used to refute the theory.

All these arguments show that RMV does not hold water. Nevertheless, this does not suffice. We shall analyze all the arguments of RMV. There are three main arguments: the change of meaning of the terms, the use of untranslatable languages, the theory-ladenness of observations. We shall examine them in the above-mentioned order.

5.2. THE PROBLEM OF MEANING VARIANCE

Let us consider Kuhn's classic example. He says that the term 'mass' has different meanings in CM and in STR (the same example is used also by Feyerabend): the Newtonian mass is stable, independent of the velocity, whereas the Einsteinian one depends on the velocity; the latter is equivalent to energy in contradistinction to the former, etc. The thesis that the concept of mass (and many others) changes its meaning in STR is acknowledged by many authors, often by adversaries of RMV. They point out only that these changes are not radical, not big (e.g. Kordig, 1971). This does not, however, explain the problem. We shall examine it more thoroughly.

I will not discuss the problem of meaning more extensively, but will confine myself to the main issue. I do not understand the alleged 'dialectical' view that the whole theory, the whole context determines the meaning of any of the employed words. On the contrary, in order to understand a theory we must know the meaning of terms it contains (cf. Achinstein, 1964). The meaning of a word is determined by its definition.[1] When the definition of an object changes, the meaning also changes. When the definition is unchanged, the meaning is also unchanged, even when we rapidly change our opinion about this object. We then formulate some new synthetic statements about it, but no analytic ones.

Let us return to the concept of mass. CM defines mass as a measure of inertia (initially Newton defined mass as quantity of matter but this definition was later given up); STR does the same. In CM there is also an alternative definition: mass is the gravitational charge (the gravitational mass is considered as equal to the inertial one); this definition is also admitted in STR (moreover, Einstein justified theoretically the identity of both masses). Hence, the meaning of the term 'mass' is the same in CM and in STR.

As is known, CM considers mass to be constant, STR considers it to be dependent on the velocity. We may present this as follows:

(5.2.1) CM: $m(x) = $ const.

(5.2.2) STR: $m(x) = f[v(x)] \neq$ const.

The statements (5.2.1) and (5.2.2) are incompatible but they use the same concepts (otherwise, they would be not incompatible).[2]

The Kuhnian thesis, that the 'Newtonian' mass is constant and the 'Einsteinian' one is variable, is misleading. There are not two concepts of mass. There is only one concept of (complete) mass. Newton and Einstein have formulated incompatible statements about it. Newton's statement (5.2.1) is false, Einstein's (5.2.2) is true. The situation seems to be quite trivial.

The adherents of RMV give us also another, apparently stronger, argument. They say that the concept of mass has been changed not only semantically but syntactically as well. In CM mass was a function of one variable (body), in STR it is a function of two variables (body and its velocity). This may be presented as follows:

(5.2.3) CM: $m = f(x)$

(5.2.4) STR: $m = f(x,v)$

This argument was put forward by Philipp Frank (cf. Giedymin, 1973) and used (usually without mentioning Frank) by other authors (e.g. Feyerabend, 1962a). However, this argument is not convincing, either. When a biologist observes the development of a young animal or plant and weighs it in different time moments, he treats its mass as a function of time, i.e. as a function of two variables:

(5.2.5) Biol.: $m = f(x,t)$.

We have an alternation of syntax here. Does this mean that the biologist uses a concept different from the physical one? After all, not only biologists do this. Astronomers also investigate the change of the mass of a star in the course of its development (as a result of the flight of matter from its surface, etc.). Does this mean that their concept of mass, though identical with the biological one, is different from the physical one? We have arrived at the absurdity. Hence, our assumptions must be false.

The statement that the mass of an organism changes in the course of its growth is a *synthetic* one and does not affect the meaning of the term 'mass'.

We may say the same about the statement that mass depends on velocity. To be sure, the former statement always seemed natural and nobody supposed that it was connected with an alternation of meaning, whereas the latter one was a big surprise and a sign of revolution in physics which could suggest a meaning variance. However, both statements are in an identical situation from the logical point of view.

We can consider another classic example of scientific revolution: Bohr's quantum theory of radiation. It gave a new formula for the frequency of radiation of an electron in the atom's orbit. Many authors distinguish the 'classical frequency' and the 'quantum frequency' symbolizing them by different indices, e.g. ν_{cl} and ν_q. It is perhaps convenient but misleading: it gives the impression that we have two different concepts of frequency. However, Bohr did not alter the concept of frequency. Its definition is the same in CM and in QM: frequency of radiation is the number of oscillations in a unit of time. Hence, again we have new synthetic statements about the frequency of radiation but no analytic ones. The meaning of this term has not changed in the course of scientific revolution.

This does not mean, of course, that meaning never changes. Often scientists use a new definition of a term, and its meaning is thereby altered. However, they always take the old definition into account.

Sometimes a new definition alters the intension but keeps the extension of the concept. For example, after Robert Koch's discovery of tuberculosis bacilli physicians passed from the definition that tuberculosis is a disease which creates tubercula in the lungs to the definition that tuberculosis is a disease caused by Koch bacilli. The extension remains unchanged, hence all statements about tuberculosis have preserved their truth-value after the change of definition. Some analytic statements have just become synthetic (e.g., tuberculosis causes changes in the lungs) and some synthetic ones have become analytic (e.g., tuberculosis is caused by bacilli).

In other cases the extension alters: it becomes wider or narrower. Then scientists search for the relationship of the old and the new concept; they state a correspondence between them. They never say: from today on we use this term in a completely different meaning. In these cases, after the change of the definition, not all statements preserve their truth-value. Some of the laws from the previous period lose their validity in the new period; they need a qualification, etc.

5.3. THE PROBLEM OF 'UNTRANSLATABLE' LANGUAGES

Adherents of RMV often say that not only the meaning of terms changes but that the rival theories use untranslatable languages; in any case, it is impossible to translate adequately the one into another (Kuhn, 1962; Feyerabend, 1962a; Szumilewicz, 1974). Karl Popper calls this view 'the Myth of the Framework', because, according to it, each theory (paradigm) creates its own framework which cannot be critically discussed; there is no communication between different frameworks. This myth is a bulwark of relativism and irrationalism (Popper, 1970).

I would go even further. Does the problem of translation arise at all in each transition from one paradigm to another? Consider again the classic example of CM and STR. What should we translate? As we have seen, both theories use the same concept of mass, formulating incompatible views about it. We may say the same about other concepts, e.g. the concept of distance. According to CM, distance is independent of the coordinate system; according to STR, it depends on the coordinate system. We have, again, two incompatible statements expressed in the same language. They use the same concept of distance − otherwise, these two statements could not be incompatible.

Certainly, the languages of CM and STR are not identical: the latter has a richer vocabulary. It contains, for example, two concepts of mass: rest mass and complete mass (the sum of the rest mass and the relativistic mass). It is often said that we deal here with a 'splitting' of the concept. We may use this expression and even treat it as a dialectical process, but we must remember that only one of these new concepts has the same definition, therefore the same meaning, as the old one. *Whole mass* in STR has the same meaning as *mass* in CM: the measure of inertia. *Rest mass* has in STR a similar definition except for one qualification: in the system in which it rests. We can add here that the last concept is also defined in the language of CM: the words 'system' and 'rest' belong, of course, to its vocabulary. Again, there is nothing to translate.

In the case of QM, it is true that the situation is more complicated. QM contains terms which are not definable in the terms of CM, e.g. 'spin' or 'strangeness'. However, I do not understand why we should try to translate them into the language of CM. For the aim of comparison of the experimental results of both theories? But CM does not predict any values of spin or strangeness, hence there is nothing to compare. The confrontation of two theories is meaningful only in the event they both predict some values of a magnitude. Sometimes it is the same magnitude, identically defined, as was the case with the instance of frequency in CM and QM. Then there is nothing to translate, we must just compare the values. In other cases the definitions are different. An example is momentum, which is defined in QM as an operator. But there is a correspondence between classical and quantum concepts here. Physicists have no difficulties with the translation of one into another; otherwise, the experimental confrontation of both theories would be impossible. The problem of 'untranslatable languages' does not exist in science.

Sometimes in physics the problem of translation is indeed important and not trivial. Namely, when physicists want to show the identity in matter (the isomorphism) of two theories which had been formulated in different mathematical apparatuses. A classic instance of this situation is presented by Schrödinger's wave mechanics and Heisenberg's matrix mechanics. As Schrödinger has shown, their subject matter is identical, hence we speak about two mathematical forms of one theory — QM. Some authors now call into question the identity of Schrödinger's and Heisenberg's theories (cf. Bunge, 1970b) but this is a different matter altogether. In any case, the task of translation is a meaningful one only when we want to express the same theory (the same statements) in another language.

5.4 THE PROBLEM OF THE 'THEORY-LADENNESS' OF FACTS

Adherents of RMV criticize the dichotomic division of scientific terms into *observational* and *theoretical* ones. They stress the theoretical impact of all observational concepts, their *theory-ladenness*. Therefore, they are sometimes called *omnitheoreticians* in opposition to adherents of a *two-level picture* of science, such as Hempel or Feigl (cf. Tuomela, 1973a).

The criticism of the two-level picture is to a large extent justified. There are no bare facts, no pure observations; reasoning and concepts intervene in every experience. But adherents of RMV go further: they say that each observation is determined by a theory and therefore two observers admitting different theories see different things.

N.R. Hanson said that Tycho de Brahe and Kepler saw different images when they observed the sunrise. The former, as a geocentrist, saw the Sun rise upon the Earth; the latter, a heliocentrist, saw the horizon lower and expose the place where he stood to the Sun's radiation (Hanson, 1958, Ch. I). But could Kepler indeed see it? We all are heliocentrists now but when we observe the sunrise we always see the Sun rising. Our perceptions are geocentric and probably cannot be different. We do not see the movement of the Earth, we *know* about it. Tycho and Kepler interpreted the same fact in different ways, their opinions were incompatible. If they saw different things, there could not be a contradiction between their opinions, because they would interpret different facts.

Nevertheless, as we have noticed, there are no bare facts, no pure perceptions, as former positivists believed. All facts are theory-laden, or rather 'concept-laden'. When we observe, we always *think* at the same time, we use (often invulnerably, spontaneously so) the concepts classifying objects and events which we observe. When we observe the sunrise we do not say that we see a shining yellow circle; we say that we see the Sun, i.e. a shining celestial body, distant from the Earth, etc. These concepts were, however, common to Tycho and Kepler; they both used them, hence their perceptions were similar.

Not always do two observers see the same thing. Where a man from our civilization sees a refrigerator, an aboriginal from New Guinea, who has not yet seen refrigerators, notices a large white box. Where a microbiologist sees bacteria, an outsider notices only a dark spot. But when adherents of two rival theories dispute, they see the same facts, the same things, although they give different and often incompatible interpretations of them. In the cases mentioned above the situation was different. We do not claim that the refrigerator is not a white box and the aboriginal from New Guinea would not claim that the box is not a refrigerator. There is no common language here and there is no contention. In science there is always a common language and there is a contention.

Stefan Amsterdamski distinguishes between a 'phenomenon of Nature' and 'a fact'. He gives the following example. Aristotle, Galileo, Newton and Einstein in describing a falling body dealt with the same phenomenon of Nature but with different facts. For Aristotle the fall of a body was its aiming at natural place, for Galileo a movement in infinite isotropic space (without any aim), for Newton a movement under the influence of gravitational force, for Einstein a movement along a geodetic incurved Riemannian space (Amsterdamski, 1973, Ch. 7, §2). However, all these differences are not of a factual but of a theoretical nature. There are, certainly, deep disagreements among the four scholars' interpretations of fall mentioned by Amsterdamski. Some of their theses are incompatible (e.g., some theses of Aristotle and Galileo), others are compatible and complementary (e.g. some theses of Galileo and Newton) but we are not dealing with a dispute about the facts in this case. Of course, the observation of a falling stone is not a 'pure perception': we do not see a moving spot but a stone moving towards the Earth's surface. It was, however, a perception common to Aristotle, Galileo, Newton and Einstein. Only their theoretical – and, of course, synthetical – statements about the trajectory of this movement and its cause are different.

Each fact is 'impregnated' by some knowledge, always by a common sense knowledge, often also by a scientific knowledge (which gradually becomes a common sense knowledge). E.g., we usually consider the flow of electric current through a conductor to be a fact although we do not see the current, we can only see its effects (thermal, magnetic, chemical, etc.). The statement that current flows, when some effects are present, is usually unquestionable and therefore we consider current flow to be a fact. If, however, somebody calls into question the statement that current flows through a given conductor, we immediately suspend our view that current flow is a fact. We go back closer to the perceptions and find a phenomenon which is admitted by all observers, e.g. deflection of the pointer of a galvanometer. We consider this deflection as a fact and current flow as an interpretation. Of course, we never reach 'pure' perceptions, we never say that the fact consists of a movement of something dark and long against a white background, etc. There is always a theoretical ingredient in the facts but we may increase or decrease this ingredient, depending on circumstances.

To put it briefly, in the event of a competition between two theories, scientists consider these descriptions of experience to be facts which are neutral with respect to the theories and shared by the competing sides. Only then the discussion has a theoretical nature; it is not a 'dispute about facts'. The latter also happens in science, namely when two observations give incompatible reports. There are different possibilities here as to the explanation of this situation: maybe an experimental device is spoiled, an observer suffers from hallucinations, he cheats, etc. These cases are not very interesting in the theoretical respect. Competition among theories, e.g. between Aristotle's and Galileo's mechanics or between Newton's and Einstein's, is always finally decided by the facts which are neutral with respect to the competing theories.

All facts are more or less theory-laden. There are no bare facts apart from all theories. However, there are neutral facts with respect to *some* theories and we can always find the neutral facts for any two rival theories (cf. McLaughlin, 1971).

Kuhn and Feyerabend store all of our knowledge in meaning. Then the difference between analytic and synthetic statements ceases to hold and logical analysis of knowledge is impossible (cf. Putnam, 1965). Hanson and Amsterdamski store all of our knowledge in perceptions, in facts. Then the difference between experience and theory ceases. As we have seen, both differences are relative. However, their liquidation precludes any analysis of knowledge.

5.5. VARIOUS RELATIONS AMONG THEORIES

In order to examine different relations among theories more carefully, we shall use the concepts of their domains and their vocabularies.

The domain D of a theory T is the set of objects (i.e. things considered with respect to some of their properties) described by T. The vocabulary V of T is a set of empirical, mathematical and logical terms used in the formulation of T. A vocabulary is, of course, only part of the language of a theory but other parts of the language (e.g., the syntax) are irrelevant here, hence we may confine ourselves to V.

Two theories T_1 and T_2 are incommensurable (incomparable with respect to their logical values) when their domains have no common elements:

$$(5.5.1) \quad D_1 \cap D_2 = \varnothing.$$

Example: the Heliocentric Theory (HC) in astronomy and QM.

The vocabularies of T_1 and T_2 are then different although they usually contain common elements (e.g., common mathematical and logical terms):

$$(5.5.2) \quad V_1 \neq V_2.$$

Consider now the case when T_1 and T_2 have common objects, i.e.

$$(5.5.3) \quad D_1 \cap D_2 \neq \varnothing.$$

Then T_1 and T_2 are commensurable because they may be compared against these common objects. There are two possibilities here: either T_1 and T_2 are compatible or they are incompatible (contradictory).

We begin with compatibility. We shall consider two important forms of compatibility: identity and reduction.

Two theories can be *identical* in matter despite a difference in form. Examples: (1) the matrix and the wave form of QM; (2) different mathematical forms of CM.[3] In this case the domains of T_1 and T_2 are identical:

(5.5.4) $D_1 = D_2$.

The vocabularies, however, are different:

(5.5.5) $V_1 \neq V_2$.

The identity in matter means that a translation of T_1 and T_2 and *vice versa* is possible. We designate by $V_2(T_1)$ the translation of T_1 into the language of T_2, the opposite translation by $V_1(T_2)$. In our case $V_2(T_1) = T_2$ and $V_1(T_2) = T_1$. A double translation is possible when a one-to-one correspondence exists between both vocabularies:

(5.5.6) $V_1 \leftrightarrow V_2$.

Of course, this relation is fulfilled only in the ideal case. In real physical theories its reconstruction is probably impossible. This impossibility encourages the questioning, mentioned above, of the identity of two forms of QM.

Let us consider *reduction* now. We shall confine ourselves to the case of microreduction, i.e. of the reduction of a theory of a system T_1 to a theory of its elements T_2. We may consider in this case the domain D_1 of the reducendum T_1 as a subset of the domain D_2 of the reducens T_2:

(5.5.7) $D_1 \subset D_2$.

Examples: (1) KL deals with an astronomical system which is a special case of a mechanical system considered by CM, i.e. the domain of KL is a subset of the domain of CM. (2) CSM deals with any large set of particles, TD deals with a portion of a gas which is a large set of molecules, i.e. the domain of TD is a subset of the domain of CSM.[4]

As we know, the first example is an instance of homogeneous reduction, the second one of heterogeneous reduction. In the first case the languages are basically the same: the structure S of an astronomical system is formulated in the terms of CM. However, not all the terms of CM are used in KL. Hence, V_1 is here a subset of V_2:

(5.5.8) $V_1 \subset V_2$.

In the second case the languages are different and we must make a translation by means of bridge rules B. All terms of TD are translated into the vocabulary of CSM, but not all terms of CSM may be translated into the vocabulary of TD. Hence, there is a one-sided correspondence between V_1 and V_2:

(5.5.9) $V_1 \rightarrow V_2$.

The relations between the theories in both cases have been presented in the Ch. 3 ((3.5.7) and (3.2.1)).

We shall pass to *incompatibility* now. Its simplest case may be presented as follows. T_1 and T_2 have the same domains and the same vocabularies:

(5.5.10) $D_1 = D_2 \qquad V_1 = V_2$.

Some main theses of T_1 entail negations of the theses of T_2 and *vice versa*. When we consider a theory as a conjunction of its statements, we may say that from each theory the negation of the other follows:

(5.5.11) $T_1 \Rightarrow \neg T_2 \qquad T_2 \Rightarrow \neg T_1$.

Example: the heliocentric theory of Copernicus (HC) and the geocentric theory of Ptolomy (GC).

We shall call this relation a *simple contradiction*.

The CR analyzed in Ch. 4 is a more interesting case. As we know, the corresponded theory T_1 is false in the light of the corresponding theory T_2, hence, we have a relation of incompatibility here. However, a theory T_1' (reinterpreted T_1) turns out to be true (as an idealizational theory), hence there is a compatibility, in a special sense, between T_1 and T_2. Therefore, the correspondence relation is not a simple contradiction but rather a *dialectical* one.

What are in this case the relations between domains and vocabularies?

When T_1 is formulated, its alleged domain D_1 is the same as the domain D_2 of the theory T_2 formulated later. However, in the light of T_2 the real domain D_1 of T_1 is empty; there is only an ideal domain D_1' of T_1'. However, it will be more convenient to use the approximative version of CR here. Then the domain D_1, in which T_1 (considered as a factual theory) is approximately fulfilled, is a subset of D_2:

(5.5.12) $D_1 \subset D_2$.

Example: CM is approximately fulfilled for all bodies which move with a velocity far from the light velocity ($v \ll c$).

Usually T_2 uses the terms of T_1 and some additional terms, hence

(5.5.13) $V_1 \subset V_2$.

Example: in STR an additional term such as 'rest mass' is used. The relation between the theories has been presented in Ch. 4 (4.3.4).

In some other cases of CR the situation is slightly more complicated. Let us take the example of the relation between QM and CM. The relation between domains is the same here: CM, taken as a factual theory, is fulfilled with a good approximation in the domain D_1 (in which h may be neglected) which is a subset of the domain D_2 of QM. However, the relation between vocabularies is not so simple. Many basic concepts of QM have new definitions (they are interpreted as operators). Each concept of CM has a corresponding one in QM. Besides, there are additional concepts in QM. Therefore, there is a one-sided correspondence between the vocabulary V_1 of CM and the vocabulary V_2 of QM:

(5.5.14) $V_1 \rightarrow V_2$.

The relation between the theories is the same here as in the previous case.

We have, therefore, two forms of CR. The typical instance of the first form is the relation between STR and CM, the typical instance of the second form is the relation between QM and CM. We may call the first form a *homogeneous correspondence* (the same language, no need of translation), and the second one a *heterogeneous correspondence* (different languages, a need of translation).

All relations considered above may be presented in the following table (which does not give an exhaustive classification of theory relations).

No.	Name	Domains	Languages	Theories	Examples
1	Incommensurability	$D_1 \cap D_2 = \emptyset$	$V_1 \neq V_2$	$T_1 \neq T_2$	HC and QM
2	Commensurability	$D_1 \cap D_2 \neq \emptyset$			
2.1	Compatibility				
2.1.1	Identity of Subject Matter	$D_1 = D_2$	$V_1 \leftrightarrow V_2$	$\begin{cases} T_2 = L_2(T_1) \\ T_1 = L_1(T_2) \end{cases}$	Two forms of QM
2.1.2	Reduction				
2.1.2.1	Homogeneous	$D_1 \subset D_2$	$V_1 \subset V_2$	$T_2 \wedge S \Rightarrow T_1$	KL and CM
2.1.2.2	Heterogeneous	$D_1 \subset D_2$	$V_1 \rightarrow V_2$	$T_2 \wedge S \wedge B \Rightarrow T_1$	TD and CSM
2.2	Incompatibility				
2.2.1	Simple Contradiction	$D_1 = D_2$	$V_1 = V_2$	$\begin{cases} T_1 \Rightarrow \neg T_2 \\ T_2 \Rightarrow \neg T_1 \end{cases}$	GC and HC
2.2.2	Correspondence				
2.2.2.1	Homogeneous	$D_1 \subset D_2$	$V_1 = V_2$	$\begin{cases} T_2 \Rightarrow aT_1, \text{ in } D_1 \\ T_2 \Rightarrow \neg aT_1, \text{ in } D_2 - D_1 \end{cases}$	CM and STR
2.2.2.2	Heterogeneous	$D_1 \subset D_2$	$V_1 \rightarrow V_2$	$\begin{cases} T_2 \Rightarrow aT_1, \text{ in } D_1 \\ T_2 \Rightarrow \neg aT_1, \text{ in } D_2 - D_1 \end{cases}$	CM and QM

NOTES TO CHAPTER 5

[1] To be sure, the definition of a term is not always explicitly formulated in science. If not, we must reconstruct it, examining the use of the term by the scientists. Sometimes we consider a set of meaning postulates rather than a definition.

[2] We may add that, according to RMV, the concept of velocity changes during the transition from CM to STR, too. The sentences 'mass is independent of the velocity' (CM) and 'mass depends on the velocity' (STR) would not contradict each other from this point of view (cf. Fine, 1967).

[3] Mario Bunge distinguishes the isomorphism of theories (giving the example of two forms of QM) and the equivalence of theories (giving the example of different mathematical presentations of CM) (Bunge, 1970b). We do not distinguish these cases.

[4] To be sure, *different* aspects of a gas as a material system are objects for these theories: TD considers a portion of a gas as a whole (macroscopic), CSM considers it as a specific set of molecules (microscopic). This difference is often important but we may disregard it here.

CHAPTER 6

THE TYPES OF METHODOLOGICAL EMPIRICISM

6.1. INDUCTIVISM

There are different views about the methods of empirical justification of laws (hypotheses) in science — different kinds of methodological empiricism.

The first and simplest one is *inductivism*. It was founded in England in the seventeenth century by Fr. Bacon and further elaborated in the nineteenth century by J. Herschell and J.St. Mill. According to this view, different kinds of inductive reasoning (in the first line — eliminative induction) provide valid knowledge. Induction is considered both as a heuristic method of investigation and a justification of its results. A law of Nature discovered by inductive reasoning based on a large set of facts is held to be true. Therefore, this view is also called *justificationism* or *verificationism*.

Inductivism was criticized already in the seventeenth to nineteenth centuries from different points of view.

Skeptical empiricists (D. Hume) revealed the fact that induction does not justify its results. However, they did not see any other way for such a justification.

Rationalists (apriorists) — from Descartes, Spinoza, Leibniz, Kant, Schelling, Hegel to Neo-Kantists and Neo-Hegelists — stressed the role of deduction, saw the way for justification of laws in aprioristic reasoning, but usually did not analyze the methods used in empirical sciences. Therefore, they did not seriously affect these sciences.

A common fault of both inductivism and apriorism was the claim that knowledge ought to be certain, thus disregarding — or, in any case, underestimating — hypotheses, approximation and probability in knowledge.[1]

Classics of dialectical materialism (F. Engels) pointed out the unity of induction and deduction, the great role of hypothesis and idealization in science, and the role of relative (approximate) truth, but did not develop these ideas.

Some British methodologists (W. Whewell, W.S. Jevons) analyzed the history of natural sciences and pointed out that their main achievements were not due to induction. They stressed the role of hypotheses, of creativity and intuition in the course of discovery (without falling into apriorism or irrationalism).[2] However, these foreshadowing ideas were scarcely noticed. The domination of inductivism among scientists – especially in Great Britain – was so strong in the nineteenth century that even those who were in doubt about the basic role of induction, like Faraday, did not venture to express their doubts in public (cf. Williams, 1968). Exceptions (Liebig, Claude Bernard) were rare.

In the beginning of the twentieth century the conventionalists (H. Poincaré, E. Le Roy, P. Duhem) revealed – albeit with exaggeration – the conventional constituents of science, the difference between scientific facts and bare perceptions (crude facts) and the methodological difficulties this gave rise to.

In spite of all these arguments, inductivism was continued by neopositivists in the period between the two World Wars, with respect to empirical sciences about Nature and society (mathematics and logic were considered by them – in contradistinction to Mill – as being purely analytical and saying nothing about the external world). However, even in this domain inductivism was essentially modified. The members and adherents of the Vienna Circle were not interested in the methods of discovering laws, of creating theories, of the growth of science. They made a sharp distinction between the context of discovery and the context of justification (H. Reichenbach's expressions). The first was relegated to the psychology of research and only the second one was held to be the proper subject for the philosophy of science. The task of the latter was the analysis of language and of the ways of justification of theses (laws) formulated in science. The methods for obtaining these theses were not examined. Induction was still considered as the method of all empirical sciences but the meaning of the term 'induction' gradually widened. It was considered as the method for non-demonstrative inference of a general statement from facts, not only in a direct way. More and more often general statements were considered as hypotheses which are only probabilistically justified by facts, 'confirmed' by them. In other words, the neopositivists (or rather post-neopositivists) gradually passed from simple inductivism (justificationism) to conformationism and hypothetism, which will be the subject of the next section.

Nevertheless, even today many scientists and philosophers are simple inductivists. Scientists rarely work by means of enumerative or eliminative induction but they often believe they are doing so in every case. There are also philosophers who still hold enumerative induction to be the main method of empirical sciences. Some Marxists interpret Engels' thesis about the unity of induction and deduction and Marx' thesis about the concrete and the abstract in a very simple, 'Baconian' way: a scientist passes at the beginning from facts (concrete) to generalizations (abstract) by means of induction and afterwards from general statements back to facts by means of deduction (e.g. Rozental, 1968). They are sharply criticized by other Marxists (cf. Nowak, 1971, 1975).

6.2. HYPOTHETISM

The second kind of methodological empiricism gives the following description of the basic method of empirical sciences. When scientists want to explain a set of phenomena they create a hypothesis (a general statement about the causes of these phenomena, etc.) which is a candidate for becoming a law. Next they deduce different, empirically testable, consequences from it and test them in experiments. If the tests give positive results, the candidate is admitted as a law, though this decision is never final. If the tests give negative results the candidate is rejected.

This view may be called *hypothetism* because, according to it, the investigation starts from a hypothesis. And there are no rules for creation of hypotheses — invention and intuition play a crucial role here. This view is also sometimes called *deductivism* because, according to it, not induction but deduction plays a crucial role in the discovery of laws. Nevertheless, of course, it is a kind of methodological empiricism because it holds experimental tests to be the device for judging the hypotheses.

We may consider Whewell, Jevons, Engels as forerunners of hypothetism. Its main creator and propagator is Karl Popper who delivered a general criticism of the Vienna Circle's inductivism in his *Logik der Forschung* and proposed the term 'deductivism' to designate his own view.[3] Popper stresses the procedure of falsification. According to him, scientists ought to take into account less probable hypotheses at first and expose them to the most severe tests. If a hypothesis is falsified it must be rejected. If it is not falsified it is 'corroborated' and temporarily persists, but this does not mean that it is true or even probably true; further tests sooner or later usually falsify it. The falsification is complete, conclusive, the corroboration is never so

(Popper, 1935). Therefore, Popper's view is also often called *falsificationism*.

Many other analytic philosophers, among them former members and adherents of the Vienna Circle (H. Reichenbach, R. Carnap, E. Nagel, C. Hempel, H. Feigl) admitted the main ideas of hypothetism. They consider theories of empirical sciences as hypothetico-deductive systems. But they usually prefer to stress the positive, though not conclusive, results of experiments. They speak not only about the falsification of statements which do not pass the tests but — first of all — about the confirmation (justification with a probability) of the ones which do pass the tests. They do not recommend beginning with less probable hypotheses. Hence, they admit the idea of hypothetism (deductionism) although not of falsificationism.

In other words, I consider *falsificationism* and *confirmationism* to be two variants of *hypothetism*. The strife between falsificationists (Popperians) and confirmationists (Carnapians) is often bitter. Nevertheless, I think the difference between them is not very essential. Both admit the same scheme of testification: *hypothesis — deduction — experiment*, i.e. the hypothetico-deductive method. The differences are rather in accents: the former stress the negative results of tests, and the latter the positive results.

There are also differences in the classification of methods. Falsificationists oppose the hypothetico-deductive method to the inductive one, and claim that the inductive method is not used in advanced sciences. Confirmationists usually consider the hypothetico-deductive method as a kind of inductive method. Some of them add qualifications and speak about the 'explanative induction' (Reichenbach, 1951; Ch. 14), 'induction in a wider sense' (Hempel, 1966, Ch. 2), 'inductive way of reasoning' (Ajdukiewicz, 1965, §52). Others use this word without qualifications and say simply that induction is a process of construction and selection of theories (testification of hypotheses) (e.g., Kemeny, 1959, Ch. 5). They all stress that here we deal with a fallible, non-demonstrative (non-conclusive) method of reasoning. This is undoubtedly true. Nevertheless, it seems that the name 'induction', even with qualifications, may be misleading here and that it is superfluous. We consider the hypothetico-deductive method to be a separate kind of fallible reasoning, apart from different variants of induction, analogy, etc.

Some authors stress the fact that we start not from hypotheses but from facts which we have to explain (Reichenbach, 1951; Hanson, 1958). This is undoubtedly true. Since the hypothesis aims to explain some facts, we must describe these facts and then search for explanatory hypotheses. This is so obvious that it would scarcely be questioned by anybody. Maybe the fault of many hypothetists (falsificationists) is the failure to say this explicitly.

Classical hypothetism separates discovery and testification. Since there are no rules for the process of creating a hypothesis, this process is relegated to the psychology of research. Hence, Popper, in his early works in any case, deals only with the context of justification (falsification is a negative justification).[4] We can see an essential difference between classical inductivism and hypothetism in this. However, as we shall see, the further development of Popperian hypothetism has led to also taking into account, in some measure, the context of discovery. In this respect the evolution of falsificationism takes an opposite direction to the evolution of inductivism.

Classical inductivism and classical hypothetism have, however, a common shortcoming. They consider only relations between a law (a hypothesis, a theory) and facts. They do not consider relations among different theories (hypotheses).

6.3. PLURALISTIC HYPOTHETISM

In the 60's Kuhn and Feyerabend delivered a general attack against both inductivism and deductivism, blaming them for neglecting the real history of science. Kuhn especially criticized the Popperian thesis that a theory is rejected when it is falsified by a fact. He showed that each theory meets with different anomalies, i.e. facts contradicting it, but is never rejected by scientists unless they find a new theory better than the previous one (i.e. being a better instrument for 'puzzle-solving' – in the Kuhnian intrumentalistic approach). There is competition among theories in a period of crisis and at the beginning of a scientific revolution. In the issue of the revolution a new theory gains and an old one is eliminated. Both verificationism and falsificationism are wrong: admittance of a new theory and failure of an old one is a joined process of verification and falsification (Kuhn, 1962). Feyerabend stresses the competition among theories even more strongly; he sees it in the whole history of science. He points out that refutation of a theory is no more final than an approbation: sometimes a theory rejected in the past returns later into science in a new form (Feyerabend, 1962a,b).

Kuhn and Feyerabend have come to relativistic and anarchistic conclusions. They do not speak about any rules of justification (or rejection) of hypotheses, hence, they do not propose any new kind of methodological empiricism. Nevertheless, their criticism was fruitful, and their idea of the pluralism of competition among theories is very valuable. This idea deeply influenced contemporary philosophy of science.

The activity of Kuhn and other critics caused Popper and his adherents to make some corrections in their views. These corrections have strengthened their position and enabled them to refute relativism in a more efficient way. We see an interesting analogy to the Counter Reformation here, which was able to efficiently combat the Reformation only by assimilating some of the opponents' ideas (cf. Kołakowski, 1962; see also Amsterdamski, 1973, Ch. 6).

A renewal of Popperian hypothetism was delivered mainly by Imre Lakatos. He admitted that scientists do not reject a theory only for reason of its contradiction with some experiments. Moreover, a new theory usually floats in 'an ocean of anomalies'. The necessary condition for the rejection of an old theory is the existence of a new, better one. Lakatos introduced new concepts: rival 'scientific research programs', distinction of their 'hard cores' and 'protective belts' of auxiliary hypotheses, negative and positive heuristics, theoretically and empirically progressive programs, etc. (Lakatos, 1970). The Lakatos conception tries to take the real history of science into account, which is a much more complex task than the original Popperian scheme supposed. In this respect Kuhn's criticism was useful for the Popper school although basic differences remain.

Lakatos called his conception a 'sophisticated falsificationism' (Lakatos, 1970). However, this name is not a very appropriate one. The idea of falsification is not so distinctly expressed here as in the original Popperian conception. Lakatos proposes not only a 'negative heuristics' but also a 'positive' one, he speaks not only about the refutation of a degenerative program but also about the preservation of a progressive program, about the means for this preservation (protective belt), etc. And failure of an old program is possible only in connection with victory of a new one. We have here destruction and construction at the same time (cf. Pietruska, 1975, p.58).[5] Later Lakatos abandoned the name 'falsificationism' himself and called his conception a *methodology of scientific research programs* (Lakatos, 1971). I think the best short name for it is *pluralistic hypothetism* because the main idea is competition among different hypotheses (the difference between hypothesis and research program is not very important). The Popperian conception may be called *dichotomic hypothetism* because it takes into account only two results of each test: a hypothesis is falsified or it is not.

Pluralistic hypothetism is now more and more widespread among philosophers of science, not only in the Lakatosian version and not only within the Popperian school.

Noretta Koertge stressed very strongly the idea of competition among different theories of research programs and even blamed Lakatos, probably not quite justly, for neglecting this idea, and for concentrating on the dialectic of conjectures and refutation *within* a research program and not on the competition among different programs (Koertge, 1971).

C. Hempel notices nowadays, mentioning Kuhn, that a theory which functioned effectively in different domains is usually rejected only when a better alternative theory appears (Hempel, 1966, Ch. 4). These examples may be multiplied.

An interesting heuristic conception has been elaborated by Asari Polikarov (Bulgaria). According to it, given a complicated problem, it is necessary to make a 'field of possible solutions' (FPS) to juxtapose all theoretically possible project-solutions (hypotheses, conceptions), and then to eliminate one by one solutions which are unsatisfactory for different reasons and finally to find the best one (Polikarov, 1966, Ch. 7; 1973, Ch. 2). Moreover, Polikarov claims that a competition among the co-existing conceptions takes place especially in advanced science. He brands as 'metaparadigm No. 1' the Kuhnian view that there are *subsequent* conceptions (paradigms) competing with each other only in the period of a revolution. He proposes 'metaparadigm No. 2': the competing of co-existing conceptions. According to him, a transition from metaparadigm No. 1 to metaparadigm No. 2 takes place in contemporary science (Polikarov, 1973, Ch. 5). Hence, according to Kuhn, the *co-existence* of different rival conceptions characterizes the initial stages of science (and short periods of revolutions); according to Feyerabend, it characterizes the whole history of science; according to Polikarov, it characterizes contemporary science. I think Kuhn is right in this case – we shall return to these problems in the next chapter (7.5).

Pluralistic hypothetism puts the relation $Hypothesis_1 - Hypothesis_2$ in the place of the *Hypothesis – Fact* relation. Of course, hypotheses are confronted by means of facts, hence, a more developed scheme of pluralistic hypothetism is $Hypothesis_1 - Facts - Hypothesis_2$.

As we know, neopositivism and especially post-neopositivistic confirm-

ationism and the original Popperian falsificationism confined themselves
to the context of justification. They assumed that methodology cannot say
anything reasonable about the context of discovery, hence this context
should be neglected by philosophy of science. Thus they have thrown out
"the baby heuristic, with the bathwater of naive inductionism" (Post, 1971,
p. 215). Kuhn and Feyerabend have rejected this separation and directed
the attention of philosophers of science to the context of discovery. Now
the conception elaborated by Lakatos and many other philosophers (e.g.
Humphreys, 1968) deals with heuristics, though to various degrees. The
problem of association of both contexts is now vividly discussed by many
authors (e.g., in Poland, Amsterdamski, 1973; Chmielecka, 1977). However, we shall not discuss it here.

6.4. IDEALIZATIONAL HYPOTHETISM

L. Nowak proposed a new scheme for empirical testification of laws (hypotheses). Every previous methodological scheme assumed that laws are
factual. However, basic laws are idealizational, hence they are incompatible
with reality and cannot be tested directly. In order to be tested, they must
be factualized. When a fact contradicts a given law *prima facie* (an expression often used in Marx' *Das Kapital*) the contradiction may be apparent: it
is possible that this fact is compatible with a factualization of this law. The
alleged idealizational law must be refuted only when the facts contradict
all factualizations.

In other words, when facts contradict a considered candidate for a law,
there are two possibilities:

(1) A factualization of the law explains the facts. The contradiction
exists only *prima facie* and the law (as an idealizational law) is upheld.

(2) No factualization of the law leads to an explanation of the facts. The
contradiction is genuine and the law is rejected.

In the first case we have actually discovered the essential factors (formulating the idealizational law) and we must search for disturbing secondary
factors (factualize this law). In the second case we have mistaken a secondary
factor for an essential one and we must still discover the genuine essential
factor, i.e. formulate a new idealizational law and then test it in the same
way.[6]

Of course, in scientific practice the decision is not so simple. When the test gives a negative result we are in doubt as to whether we have sufficiently factualized the idealizational law in question, i.e. whether we have taken a sufficient amount of secondary factors into account. Sometimes we decide to reject the candidate for a law, sometimes we try to factualize it further. Our decision is never final. The future of science may show that we have wrongly rejected a law instead of factualizing it further. Of course, the opposite decision may also turn out to be wrong, namely when new facts contradicting the law are discovered.

I call this conception *idealizational hypothetism*. It is, of course, a kind of methodological empiricism, because it also holds experience to be the judge of hypotheses.

The conception of idealizational hypothetism as presented above was elaborated by L. Nowak. However, similar ideas, as a matter of fact, have been expressed by some other philosophers of science. E.g. Mario Bunge in a paper on the confrontation of theory with experience criticizes the primitive view of Carnap and other philosophers that an isolated hypothesis is directly confronted with pure empirical evidence. Bunge points out that theories often refer not to observed but to idealized things. Hence, they cannot be tested directly. We must find an appropriate model for a theory. Then different subsidiary assumptions are necessary to obtain new forms of theories which are closer to experience. Only in this form may the theories be tested. However, this is not enough. Experimental data are usually also unfit for the confrontation. They must be translated into the language of the tested theory; here, again, additional assumptions are necessary. Confrontation is possible only when all these introductory procedures are over. Even then it is not decisive: neither agreement nor disagreement between theories and facts prepared in this way entails any final decisions (Bunge, 1970a).

I think that Bunge's transition from the original form of a theory, referring to a model, to its 'closer to experience' form is the same procedure which is called concretization by Nowak and factualization by myself. Bunge speaks explicitly about idealized objects of a theory. Hence, he represents a version of idealizational hypothetism. Moreover, Bunge provides us with a new contribution: he points out that not only the theory but also the experimental data must be transformed. In this respect Bunge's version of idealizational hypothetism is more developed than Nowak's.

The scheme of experimental testification, according to idealizational hypothetism (in Nowak's version), may be presented as follows: *Hypothesis – Factualization – Facts.*

6.5. PLURALISTIC IDEALIZATIONAL HYPOTHETISM

W. Patryas, a young representative of the Poznań methodological school, recently introduced into the Nowak conception an alternation analogous to the one which was introduced by Lakatos into the Popper conception.

Patryas criticizes generally the common 'dichotomic conception of testification', according to which the aim of a test is the establishment of logical value of the tested hypothesis. This conception is usually connected with the following postulate: reject a candidate for a law when it is falsified (in Nowak's version: when its factualization is falsified). However, a scientific law is always false in the classical sense.[7] Therefore, if scientists followed this postulate they would have to reject all laws. However, the procedure used in real science is different. When a scientist wants to explain a set of phonomena, he examines different statements as candidates for a law. Then he chooses the one which is the best, i.e. the least false one. Hence, two logical values are not enough.

To describe this procedure more precisely, Patryas elaborates a 'pluralistic conception of testification'. He introduces a concept of *degree of inadequateness* (DI). The measured value of a considered magnitude A always deviates from the value predicted by a theoretical statement (candidate for a law). These deviations are, of course, different for different values of A (different conditions, different states of affairs). The DI of a statement is the maximum of the absolute values of these deviations. Only for a true statement is the degree of inadequateness equal to zero, for the false ones it is always positive. Scientists choose the candidate with the least DI.

Patryas associates this pluralistic conception of testification with the idealizational conception of science. He formulates the following methodological postulate.

When a DI of a considered candidate for a law exceeds the DI of its rival, try to factualize your candidate. When after all trial factualizations its DI still exceeds the DI of its rival, reject your candidate (Patryas, 1976).[8] In other words, when the DI of the candidate under consideration after its factualization is smaller than the DI of its rival, the candidate is accepted and proclaimed to be a law. Of course, this decision is not final and in future our candidate may be dismissed by another one with a smaller DI.

We may call Nowak's (and Bunge's) conceptions *dichotomic idealizational hypothetism* and Patryas' *pluralistic idealizational hypothetism.*

The scheme of testification, according to pluralistic idealizational hypothetism, is as follows: *Hypothesis₁ – Factualization – Facts – Factualization – Hypothesis₂*. This scheme may be also presented in another form: *Idealizational Hypothesis₁ – Factualized Hypothesis₁ – Facts – Factualized Hypothesis₂ – Idealizational Hypothesis₂*.

6.6. A CONFRONTATION: THE DIVERSITY OF METHODS

We have considered five kinds of methodological empiricism. We shall juxtapose their schemes using the following symbols: the pointed arrow for induction, the double arrow for deduction, the symbol ⊣ for factualization (this symbol is used by Nowak). We shall make two-dimensional schemes putting the hypotheses (theories) above and the facts below.

(1) Inductivism:

Law

Facts

(2) Dichotomic Hypothetism:

Hypothesis

Facts

(3) Pluralistic Hypothetism:

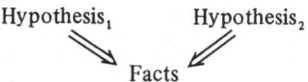

Hypothesis₁   Hypothesis₂

Facts

(4) Dichotomic Idealizational Hypothetism:

Idealizational Hypothesis

Factualized Hypothesis

Facts

(5) Pluralistic Idealizational Hypothetism:

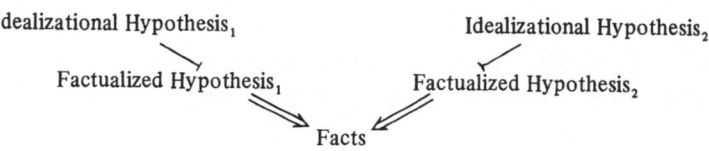

Idealizational Hypothesis₁   Idealizational Hypothesis₂

Factualized Hypothesis₁   Factualized Hypothesis₂

Facts

The first scheme intends to show the way of discovery and justification for laws, and the next four the way of justification (testification) for hypotheses (candidates for laws). The founders of every conception hold their own scheme to be the best, the most adequate with respect to the methods of science, and criticize all previous conceptions as inadequate. In each case the criticism is to a great degree justified. Nevertheless, I do not think that only one of these conceptions is the right one. On the contrary, all five methods are used in science and they are effective although their values are not equal.

(1) Induction has — in contradistinction to the opinion of Popper and Nowak[9] — its place in science, in any case in the initial stages of investigation of some kind of phenomena. Even enumerative induction enables us to discover very simple laws; e.g., each metal is a conductor of electric current, each gas is liquifiable, etc. Eliminative induction, by means of Mill's canons, often leads to the discovery of the causes of some phenomena, although only in the simplest cases. Bacon's and Mill's mistake was not the elaboration of inductive methodology but its overevaluation, the conviction that it is a unique (or main) method of empirical sciences.[10]

(2) The hypothetico-deductive method leads to more profound results. It enables us to discover more essential (factual) laws which explain some observed phenomena. E.g., the hypothesis about the existence of electrons explained well such phenomena as cathode rays, photo-effect, and it was strongly confirmed by the prediction of other phenomena discovered later in the experiment. This hypothesis had no serious rivals and was soon commonly admitted. We must stress again the fact that the statement about the existence of electrons is factual and not idealizational.

The example with the electrons is not exceptional. The method described by dichotomic hypothetism is often used in science and has led to great successes. Although Kuhn, Lakatos and Patryas are right in claiming that no hypothesis is rejected unless there is a better one, the situation is often the opposite: there is only one serious hypothesis explaining a phenomenon, hence it is admitted without competition.

(3) When different hypotheses (or research programs) are proposed in science, their competition is always fruitful as is every competition in general. The existence of rivals causes the adherents of a given hypothesis to make its examination more thorough and, maybe, to improve it (to progress within a research program). It is a strong stimulus to the growth of

science. We shall return to these problems in the next chapter.

An example of a fruitful competition between two factual hypotheses was the dispute in nineteenth century chemistry between the adherents of Prout, claiming that all atoms consist of atoms of hydrogen, hence the atomic weights are integers, and the adherents of Berzelius, revealing the fact that many atomic weights are not integers. This competition was a stimulus to a more exact measuring of atomic weights.

(4) The method of idealization and factualization enables us to discover the essential dependencies when there are different disturbing factors. Often an ideal model of a given phenomenon has no rival models. Examples are Archimedes' model of the lever and Galileo's model of fall. In these cases dichotomic idealizational hypothetism describes very well the method used in science.

(5) In other cases there is a competition among different ideal models, hence among different candidates for an idealizational law. We may take an example from contemporary physics. In the last decades different models of the structure of the atom nucleus were created: drop model, envelope model, bootstrap model. Probably no physicist claims that one of these models describes exactly the real nucleus. All these models are idealizations but their factualizations are not easy. Therefore, the problem of which model grasps the most essential dependencies is not yet solved.

The method of idealization is now widespread in different sciences, including the social sciences. In economy, psychology, and linguistics different ideal models are built and confronted, after factualization, with each other. In all these cases pluralistic idealizational hypothetism describes well the method used in science.

As we see, science uses all five methods. However, in advanced sciences the method of idealization is the basic method and competition among different hypotheses (models) is more and more common. For this reason, among the kinds of methodological empiricism described above the later ones are more adequate with respect to contemporary science than the previous ones.

The problems of testification are closely related to the problems of the development of science and the problems of truth. We shall examine them in the next two chapters.

NOTES TO CHAPTER 6

[1] To be sure, there were differences among the rationalists. E.g., in contradistinction to Descartes who neglected probable knowledge, Leibniz admitted its place in science (cf. Kuptsov, 1973, p. 81).

[2] It is noteworthy that whereas Whewell is now commonly regarded as a forerunner of modern methodology, Jevons is only rarely mentioned in this respect. In Poland, Jan Such 'discovered' him as a forerunner of hypothetism (Such, 1969).

[3] This name seems to be not appropriate. It suits more the Cartesian ideal of a certain deductive knowledge. The hypothetical nature of empirical knowledge, the beginning from a hypothesis in the process of search for laws, is the more important thesis of Popper, therefore the name 'hypothetism' is much better.

[4] Lakatos criticizes Carnap very sharply for neglecting the context of discovery and for eliminating the method of investigation from methodology (Lakatos, 1968, p. 326). However, he does not notice that Popper's original conception had the same shortcoming.

[5] However, the problem is not so simple. The falsification of a theory may be in a sense final, but the verification never (therefore, we speak about confirmation). Hence, though verificationism (confirmationism) and falsificationism are both one-sided, the latter is 'more right', which was overlooked by Kuhn (cf. Such, 1975, Ch. IV).

[6] This conception is contained in Nowak's first book (Nowak, 1971). I write 'factualization' instead of his 'concretization'. In a later book (Nowak, 1974a) a much more complicated conception is contained — I will not discuss it here.

[7] I should say: a quantitative law (see Ch. 8).

[8] Patryas' conception is much more complicated — only some its main ideas are presented here. I have slightly changed the terminology. Patryas uses, after Nowak, the term 'concretization'. True, he uses the term 'factualization' too, but in a restricted meaning, as a special case of concretization (the abrogation of the so-called 'ceteris paribus clause'). I don't distinguish among the kinds of factualization.

[9] In discussions which took place during the last years, Nowak has admitted that induction has its place in science.

[10] The modern methodologists — R. Carnap, Y. Bar-Hillel, J. Kemeny, J. Hintikka, G. von Wright, H.E. Kyburg, K. Szaniawski and others — elaborated new, more sophisticated forms of induction, namely different variants of statistical induction, connected often with decision theory, information theory, etc. These kinds of induction have a fortiori their place in different branches of empirical sciences. However, they must not be mixed up with the hypothetico-deductive method.

CHAPTER 7

REVOLUTIONS AND CONTINUITY

7.1. SIMPLE CUMULATIVISM
(NO REVOLUTIONS OR ONE REVOLUTION)

According to the traditional view, the development of science is a simple accretion of knowledge, cumulation of truths: each law or theory admitted by scientists remains without changes. The growth of science consists only of its enrichment, without losses. Even when a new, more general theory is formulated, the previous one preserves its truth-value. Of course, individual mistakes are always possible, some papers may be half-baked, but in modern times such mistakes are quickly corrected by the scientific community.

We shall call this view *simple cumulativism* (it is also called *extreme cumulativism*). It reigned till the beginning of our century. It was shared by philosophers of different orientations. Traditional empiricists (inductivists) held experience and induction to be an unshaken basis of science, therefore they were cumulativists. Rationalists believed that the mind creates the principles of science *a priori* and these principles in turn are an unshaken basis of science, hence, they were cumulativists, too. Mechanical materialists stressed the interaction of mind and experience and were convinced that this interaction gives trustworthy and ultimate knowledge about the material world, hence they were also cumulativists.

The simple cumulativists mostly admitted Fr. Bacon's idea about the *one* revolution in the history of human thought: the creation of modern science in the sixteenth and seventeenth centuries which was supposed to be a revolution of the intellect against medieval ignorance and superstitions. But they did not see any revolutions later on.

In the nineteenth century several ideas incompatible with simple cumulativism appeared. I mean some ideas in Hegel's philosophy and especially in dialectical materialism such as qualitative jumps, dialectical negation, relative truth concerning elements of falsity. However, the science of the nineteenth century was not ready to assimilate these ideas and did not provide a sufficient factual basis for their further elaboration.

The period of the great revolution in physics at the turn of the nineteenth and twentieth centuries provided conditions for the development of new views. Many scientists and philosophers in that period spoke about the crisis in science, about the ruin of the old physics, called radium 'a great revolutionist', etc. (e.g., Poincaré, 1906). Lenin entitled a chapter of his main philosophical book 'The newest revolution in natural science and philosophical idealism'; he considered revolutions in science to happen with natural regularity but stressed at the same time the objectivity of scientific knowledge (Lenin, 1909); however, he was not concerned much with problems of methodology and history of science.

Conventionalists analyzed the methods of science more closely. They revealed the revisibility of scientific statements containing conventional elements. Hence, they seriously shook cumulativism. It is true that a leading conventionalist, Pierre Duhem, stressed the continuity of the history of science. He re-evaluated medieval thought, especially of the Paris school in the fourteenth century, and denied the fact that Copernicus and Galileo caused a genuine revolution in science. For this reason, many authors say that he considered human thought to be completely continuous and opposed the Baconian conception of *One Revolution* with a conception of *No Revolutions* (Agassi, 1966; Rodnyj, 1973). However, Duhem pointed out that every law of physics is approximate and provisional; hence, he was not a simple cumulativist.

This does not mean that there are no simple cumulativists now. In twentieth century philosophy, neopositivistic inductivism has remained within the scope of cumulativism, though its representatives have no special interest in the history of science. Simple cumulativists are also to be found among the historians of science, even the most distinguished ones.

Derek J. de Solla Price writes that we deal with a cumulative growth of achievements in science reminiscent of a heap of bricks: every scientist adds his brick. Some false publications (e.g. on the alleged Blondlot's rays) are, according to this author, just fortuituous mistakes (De Solla Price, 1961, Ch. 5).

A. Ruppert-Hall takes over the Baconian idea of one revolution but expands it to three centuries (sixteenth to eighteenth). According to him, the beginning of the nineteenth century is a line of demarcation. Since that period the sciences have a cumulative nature (Ruppert-Hall, 1954, Introduction).

J. Agassi quotes many other examples of cumulativism among historians of science and sharply criticizes them, though he does not use the word 'cumulativism' (Agassi, 1963).

Physicists speak about a cumulative growth of science so often that quotations are superfluous.

There is, indeed, a cumulation of achievements in mature science, in spite of revolutions, but this cumulation is connected with corrections. We shall consider this question in the next sections. However, a thesis that in science (even, in the contemporary one) there is a cumulation *tout court*, without qualifications, is an anachronism.

7.2. SIMPLE ANTICUMULATIVISM (PERMANENT REVOLUTION OR OCCASIONAL REVOLUTIONS WITHOUT CONTINUITY)

The turn from the simple cumulativism was made in the philosophy of science by Popperian hypothetism. According to its original form, the main task of science is the falsification of alleged laws; after every falsification scientists set up a new candidate for a law, independent of the previous ones. In other words, Popper did not see — in any case in the early period of his activity — any continuity in science. His view is often called *Permanent Revolution* (e.g., Agassi, 1966), though, maybe, *Attempted Permanent Revolution* would be a more appropriate name (Berkson, 1971). Nevertheless, Popper was, in general, not interested in the history of science at that period, his attitude was, as a matter of fact, ahistorical. After all, he was the author of several papers criticizing historicism (1945–1961).

In the 60's Popper essentially changed his attitude in this respect. A Soviet author even says that the Popper of today would be met as an open adversary by the Popper from the 40's and 50's (Mayzel, 1975). Indeed, today Popper is a determined adherent of the historical attitude, in any case with respect to science. He speaks about the progress of knowledge towards truth, mentions CP, proposes an evolutionary epistemology, etc. At the same time he maintains his ideas of falsificationism, of the necessity of permanent criticism, etc. (Popper, 1963, 1970, 1972). Hence, he associates the idea of an attempted permanent revolution with the idea of continuity, of progress in science.

A general attack against cumulativism has been delivered by T. Kuhn and P.K. Feyerabend. As is known, Kuhn distinguishes two different kinds of periods in the history of science: periods of *normal science* in which continuity reigns and periods of *revolutions* which overthrow the old paradigm and break the continuity. The new theories are incommensurable with the old ones, the concept of truth is superfluous (Kuhn, 1962). This view is called *Occasional Revolutions* (Agassi, 1966).

Feyerabend preserves the concept of truth (of a correspondence to reality) but he claims that different theories are incommensurable, he denies the existence of the periods of normal science and of any continuity. Lately he has come to an 'anarchistic' theory of science refuting the necessity of method in scientific investigation (Feyerabend, 1970b). The name *Permanent Revolution* fits this view even better than the Popperian one.

Hence, early Popper and *a fortiori* Feyerabend see the history of science as a sequence of permanent revolutions without any continuity, Kuhn sees it as occasional revolutions without continuity. All these views may be called *simple anticumulativism* (or *extreme anticumulativism*). Later on Popper presented another view which shall be considered in the next section.

7.3. A DIALECTICAL VIEW (REVOLUTIONS AND CONTINUITY)

Both simple cumulativism and simple anticumulativism are extremities which do not hold water. They must be replaced by a synthesis which unites the oppositions and at the same time goes beyond them.[1] It may be called a *dialectical* view of the development of science.

Of course, it is just a general label. Different versions of this view are possible. As I have mentioned, this view is presented by the later Popper's papers (although, maybe, he would be astonished if he heard his approach called 'dialectical'). Lakatos also presents a version of this view. He sees revolutions in science: failure of one research program and victory of another; he sees also a continuity, a kind of progress, both within a given research program and in the transition from an old program to a new one.

The dialectical view is presented by all these scientists and philosophers who see the role of CP and at the same time notice the contradictions between the corresponded and the corresponding theories. As we know, CR may be considered to be a dialectical negation.

Philosophers of science in the U.S.S.R. have admitted the role of CP after 1948 (see Ch. 1). Some of them proposed, apart from CP, other methodological principles showing more directly the negative, anti-cumulativist aspect of the development of science. One author formulated a *Principle of Limitation* in physics: a new theory adds new postulates and shows the limits of the old one (Illarionov, 1964). Another wrote on two associated principles (borrowed from mathematics): the *Principle of Interdiction*: a new physical theory interdicts the use of an old one within a certain range of reality (negative aspect) and the *Principle of Permanence*: a new theory is a continuation of an old one (positive aspect) (Krymskij, 1965). In a later paper the Principle of Limitation is considered to be a synthesis of the Principle of Interdiction (negative aspect) and the Correspondence Principle (positive aspect) (Mamchur and Illarionov, 1973). Lately, even a *Principle of Non-Correspondence*, besides the CP, was proclaimed (Kard, 1975).

In the last decades an essential increase of interest in the problems of the development of science may be noticed. Discussions among western philosophers of science are vividly commented on in the U.S.S.R. Both Kuhn's and Lakatos' conceptions are received very favorably. Their anti-positivistic attitude is stressed and their turn to the real history of science is emphasized. In Kuhn's book some authors appreciate the idea of successive periods of normal science and revolutions, which resembles the Marxist idea of successive evolutionary and revolutionary stages in the development of both matter and society (Kedrov, 1969b).[2] Others, although admitting the great merits of Kuhn, criticize his exaggerations: mono-paradigmatism whereas two paradigms often exist simultaneously (Rodnyj, 1973), the simplified opposition of young (revolutionary) and old (conservative) scientists, whereas young scientists learn from the handbooks and lectures of the old ones (Mamchur, 1971). It is noteworthy that Soviet authors rarely mention Kuhn's relativism.

Lakatos' papers have been received mostly very favorably, sometimes enthusiastically (Shvyriov, 1971). It is pointed out that Lakatos – in contradistinction to Kuhn – sees the struggle among theories not only in the period of a crisis (Yaroshevskij, 1974). Some authors just remark that Lakatos' conception needs corrections on the basis of a more thorough examination of the history of science (Akchurin and Mamchur, 1972).

A vivid discussion of these problems has taken place among philosophers of science in Poland. They mostly acknowledge the merits of both Kuhn's and Popper's school and place themselves 'between' them. However, some Polish philosophers are closer to Popper and Lakatos (they stress the internal logic of science), others are closer to Kuhn (they stress the external factors – see Ch. 9). These differences have been revealed in the collection of papers on CP (Krajewski *et al.* (eds.) 1974) and in other publications. Some

philosophers and logicians go very far in the first direction and stress only
the destructive role of Kuhn's book: its denial of CP and of continuity in
science (Augustynek, 1974), its denial of any rational methodology in
periods of revolutions (Stanosz, 1972). Some others are rather close to
Kuhn, though not uncritical towards him (e.g., Amsterdamski, 1973;
Szumilewicz, 1974).

In spite of all these differences, almost all authors in the U.S.S.R. and
Poland (and probably in many other countries) agree that the development
of science is a complex phenomenon, that it includes both revolutions
and continuity.

This idea is also widespread among western philosophers of science
nowadays. The majority of them oppose relativism and breach of continuity
in science claimed by Kuhn and Feyerabend. Some of them directly admit
the unity of revolutions and continuity (e.g. Krüger, 1973). Others stress
especially continuity, sometimes even defend 'conservatism'. The paper by
H.R. Post mentioned above has a provoking subtitle 'In Praise of Conservative Induction'. It begins with six quotations from distinguished scientists about the continuity of science; it ends with the 'moral' that scientific
progress is linear. Nevertheless, even this author is against 'naive inductionism' and interprets induction in a wide sense (as a transition from less general
to more general theories). Declaring that every good scientist is conservative,
Post nevertheless admits that some new theories are inconsistent with old ones.
He agrees that revolutions exist but stresses the continuity, the 'guideline'
across them (Post, 1971). Therefore, his view is, as a matter of fact, dialectical
as well, although he accentuates continuity with some exaggeration.

When we speak about the dialectical view, we give only a general outline.
Many problems remain. One of them is the question whether CR is always
valid in science. Another is the question whether the division of the history
of science into evolutionary (normal) and revolutionary periods is justified.
We shall consider them now.

7.4. THE THRESHOLD OF MATURITY
(TWO KINDS OF REVOLUTIONS)

In contradistinction to Post, I do not think that CP is a general principle for
the whole history of science. It is valid only beginning with a definite period.

We must distinguish two big epochs in the history of every science. In the first epoch science is *immature*, in the second one it is *mature*. An immature science grasps wrongly the essence of its objects, a mature science does it correctly in rough outline. The examples presented below will illustrate this idea.

This distinction entails a distinction of two kinds in scientific revolutions: (1) the initial revolution, i.e. the transition of a given branch of science through the *threshold of maturity*, (2) revolutions during the development of mature science.

During revolutions of the first type our view about the essence of phenomena changes radically. There is a simple contradiction between main theses of an old and a new theory here. During revolutions of the second type our view does not change so radically, the science penetrates into the essence more deeply. There is a CR here, dialectical contradiction between the theses of an old and a new theory.

We shall quote three examples of revolutions of the first kind: in astronomy, in mechanics, in chemistry.

In astronomy the initial big revolution was made by Copernicus. The main thesis of his HC (the Earth moves round the Sun) implies a negation of the main thesis of Ptolomy's GC (the Earth rests in the center of the World). The GC does not grasp the structure of our planet system properly, the HC does so in a rough outline.

In mechanics the initial revolution was made by Galileo, Descartes and Newton. Their law of inertia (a body preserves its velocity when no force acts on it) implies a negation of the main thesis of Aristotle's mechanics (a body moves only when a motor acts). Aristotle did not grasp the essence of movement properly, Galileo did.

In chemistry the initial revolution was made by Lavoisier. The main thesis of his oxygene theory (combustion is a synthesis, a combining with oxygen) implies a negation of the main thesis of Stahl's phlogiston theory (combustion is an analysis, a discharge of phlogiston). Stahl did not grasp the essence of combustion correctly, Lavoisier did.

In all these cases there is no correspondence between the old and the new theory.[3] They are simply incompatible.

I would like to be understood correctly. The initial big revolution is the initiation of a *mature* science but not of science in general. An immature science is also science. Aristotle, Ptolomy and Stahl were great thinkers. Their theories provided scientific, although basically mistaken, explanations of phenomena. Their theories enabled correct prediction of many events. There is an essential difference between these theories and the pre-scientific views of magic, mysticism and religion.

Aristotle and Ptolomy had no need to analyze the magical and mythic views about the movement of terrestrial and celestial bodies. They neglected these views and could not do otherwise. We may say that mythic views and conceptions of early (immature) science were incommensurable; there was no contention.

The relationship between the first theories of mature science and its predecessors was quite different. Copernicus, Galileo and Lavoisier did not neglect the previous theories; on the contrary, they analyzed them carefully and discovered their essential faults. Moreover, they exploited the facts and many empirical regularities described by their predecessors. Copernicus appreciated Ptolomy's work very highly. Galileo also appreciated Aristotle, saying that if he had lived he would have admitted new ideas. Only Lavoisier was rather ill-disposed to Stahl (probably, the distance was too short) but also admitted some of his merits.

As we have seen, the *main theses* of Copernicus, Galileo and Lavoisier were *simple negations* of the main theses of their predecessors. However, the *whole theories* of the creators of mature astronomy, mechanics and chemistry were rather *dialectical negations* of previous theories, because some of their achievements have been maintained (the results of observations and some simple regularities). However, there was no CR between the new and the old theory. Hence, we do not identify CR with dialectical negation in science. The former is a special form of the latter.

Revolutions in mature science lead to theories which are in CR with their predecessors. Each revelation of idealizing assumptions L_1 and formulation of L_2 being in CR with L_1 may be called a scientific revolution. Of course, the depth and importance of such a revolution may vary to a large extent. Sometimes, a revolution deeply changes science and even our whole world-image, for example: QM, STR, GTR. Sometimes, it changes only a narrow branch of science, for example: the differential form of Ohm's law (a revolution in the theory of electric current) or van der Waals' law (a revolution in the theory of gases). In spite of a big difference in scope, the logical relations are in all these cases similar.[4]

My conception may be called: *one initial revolution* (without CR)[5] and *many occasional revolutions* (with CR).

This conception was first published in 1974 (Krajewski, 1974a).[6] E. Pietruska noticed that when a mature science is defined as a science in which there is a cumulation or correspondence, the thesis about cumulation or correspondence in it becomes analytical and unfalsifiable (Pietruska–Madej, 1977). She is right. Nevertheless, I think that the best definition of mature science is its definition by means of CR. The thesis that there is CR among successive theories of mature science becomes, indeed, analytical. However, there is another synthetical thesis: every branch of science crosses at some period the threshold of maturity. This thesis may be and must be tested by the history of science.

7.5. PERIODS OF EVOLUTION AND OF REVOLUTION

According to Kuhn, there are long periods of normal science when one paradigm reigns in a given branch of knowledge and relatively short periods of crisis and revolution when different paradigms compete with each other. This view has been criticized by many authors, especially during the London discussion in 1965 (see Lakatos and Musgrave (eds.) 1970). M. Masterman showed the large diversity of meaning of the term 'paradigm' in Kuhn's book (Masterman, 1970). K. Popper pointed out the danger of normal science with its uncritical attitude and the advantage of permanent competition (Popper, 1970). Many debaters questioned the 'monoparadigmatism' of the alleged normal science. They quoted different examples of competing, incompatible views in non-revolutionary periods: the long competition between wave optics and corpuscular optics, between atomism and continuity theory of matter (Popper, 1970), between catastrophism and uniformitarianism in geology and paleontology in the second quarter of the nineteenth century (Toulmin, 1970), among mechanical, thermodynamical and electrodynamical points of view in physics in the second third of the nineteenth century (Feyerabend, 1970). All these authors point out that competition among different theories is fruitful and desirable; on the contrary, monopoly is harmful.

I agree that Kuhn's word 'paradigm' is ambiguous and that this ambiguity has disadvantageous effects. As Amsterdamski correctly notices, often a new theory abandons one paradigm and continues another one. Einstein rejected some Newtonian principles but did not want to abandon the others, e.g. strict determinism, and was for this reason held by the creators of QM to be a dogmatist (Amsterdamski, 1973, Ch. 6). Therefore, I prefer to speak about *theories* and not about paradigms.

I accept also the postulate of criticism and of diversity of theories. Free competition is better than monopoly. Is, however, the competition among different theories really a *permanent* feature of science? It seems H.R. Post is right when he writes that at any time there is usually *one* dominant theory (Post, 1971, p. 220).

Consider some of the above-mentioned examples. The competition between Newton's corpuscular optics and Huygens' wave optics at the end of the seventeenth and the beginning of the eighteenth centuries was rather an exceptional situation. It was connected with the contention between Newtonian and Cartesian theories of matter described so vividly in Voltaire's famous letters from England. This situation was paradoxical not only for Voltaire. Newton's increasing authority soon prevailed to such an extent that the corpuscular theory became dominant in all European countries. A change came at the beginning of the nineteenth century after the discovery of the interference of optical rays by Young and Fresnel. There was a revival of the wave theory of light which gained adherents very quickly and which reigned in turn till the beginning of the twentieth century. Then the discovery of photo-effect and its explanation by Einstein by means of Planck's concept of photons led to a new corpuscular or rather wave-corpuscular theory of light. This theory has reigned since that time without rivals. Hence, the periods of competition of different theories of light (revolutionary periods) were very short with respect to the monotheoretic periods (evolutionary periods).

Feyerabend is right when he points out that in nineteenth century physics there existed three different, and incompatible in their consequences, points of view (methodological paradigms). But he admits himself that they reigned in different branches of physics and that their inconsistency was revealed only later on. Such a revelation caused a crisis in each case, led to a revolution, and to new, more general, theories.

The competition of Cuvier's catastrophism and Lyell's uniformitarianism in paleontology did not last long either. Darwin's work (and later Lyell's papers) led to a synthesis which was commonly acknowledged.

In this case, as in the case of optics, a competition of two one-sided theories led to a new theory which is a synthesis of the competing theories and at the same time their (dialectical) negation. The whole process may be called a *dialectical ascent* (cf. Koertge, 1971).

In contemporary science a confrontation of different theories and their synthesis is, probably, a general regularity (cf. Ovchinnikov, 1974). Nevertheless, it seems that mostly there is one dominant theory in each branch of science. Two different theories compete in periods of crises which are relatively short.

The distinction of two kinds of changes, hence of two kinds of periods in science, is admitted more or less explicitly by many philosophers. I. Lakatos, as we know, distinguishes the creation of successive theories within a research program and the shift of research programs. The first one may be called evolution, the second one revolution. In general, there are essential parallels between Lakatosian and Kuhnian conceptions of the growth of science, in spite of all the differences between them (cf. Amsterdamski, 1970, 1973).

In L. Nowak's approach there are also two kinds of changes: the gradual factualization of an idealizational law (revealing new secondary factors) when we deal only with *prima facie* anomalies, and the abandoning of the law when we deal with genuine anomalies. We should abandon the idealizational law only when all alleged factualizations fail. In other words, first one should be a 'conservatist' trying to solve the old law. However, such a 'conservatist' works for the future 'revolutionist' (Nowak, 1974a, p. 173).

Our conclusion is the following one. The Kuhnian idea of two kinds of periods in the history of science — or, rather, of each branch of science — is basically correct. However, the term 'normal' is unlucky. A revolution is also a normal phenomenon in science, which Kuhn has shown very well in his own books.[7] Therefore, I prefer to speak about periods of evolution and periods of revolution (preceded by a crisis) in the science.

I think that each branch of science (sometimes, even a very narrow one) has its own periods of revolutions. Of course, revolution in a basic domain of science often causes revolutions in many other ones. It is true that Kuhn presented the opposition between the two kinds of periods too extremely, which justified the criticism. I admit that even in evolutionary periods some scientists question the reigned theories and sometimes try to test them (cf. Williams, 1970). However, these attempts are rare and do not change the general situation in that given branch of science. On the other hand, in revolutionary periods the majority of scientists continue to solve their 'tiny puzzles' (Feyerabend, 1970a, p. 208). However, in these periods criticism increases rapidly and, even when it is an affair of a minority, it determines the whole situation. Hence, opposition between the two kinds of periods does occur in science, though not so sharply as Kuhn presents it.[8]

Of course, there are additional complications. Different branches affect each other. Philosophical outlook and admitted methodological norms affect them too. In short, there are here, as everywhere, different disturbing factors which obscure the image. Nevertheless, the sequence of evolutionary and revolutionary periods is an essential law in the history of science. As with every essential law — not only the laws of Nature — it is based on an idealized model.

In our time revolutions probably occur more often than previously. The periods of evolution are shorter. Therefore competition among different theories is not so rare as it was in the past. However, the differences among them are rather less than in the past: nobody maintains today that there are only particles or only waves, that there are only sudden catastrophes or no jumps at all, etc.

Periods of domination of one theory are now shorter, but they exist. The fact that they are shorter is very favorable: competition stimulates progress. But will the monotheoretic periods cease to occur at all in future? I think this is unlikely and even undesirable. Evolutionary periods are more favorable to the practical aims of science whereas revolutionary periods are more favorable to the theoretical aims. As Amsterdamski correctly observed, Popper and Kuhn are both one-sided: the former sees only the truth, he floats in the sky; the latter sees only the practice (puzzle-solving), he treads the Earth. Both separate the two aims of science which are intimately connected: the search for truth and the solution of practical tasks (Amsterdamski, 1973, Ch. 6).

Of course, acceleration in revealing truth leads to growing practical achievements. Therefore, the sky is useful to the Earth. Nevertheless, periods of stability are also necessary. Permanent revolution is desirable neither in society nor in science. Both a government and a theory need a certain time to prove their effectiveness.

I will now compare my view with the others.

I take over the Baconian idea of the initial revolution but I consider it not as a revolution against prejudices and obscurantism but as an overthrow of immature theories. At the same time I point out that a mature theory exploits many results of an immature one; in other words, I see a certain continuity here, according to the Duhemian view but without his exaggerations (i.e. his denial of revolutions).

I take over the Kuhnian idea of many occasional revolutions but I reject his incommensurability claim; I stress the continuity of mature science, in spite of revolutions, according to CP and the idea of dialectical negation. I agree with the Popperian idea of progress towards truth, with his appeal to constant criticism, with his view that revolutions are necessary for science, but I reject his claim of permanent revolution.

I agree with the idea of Feyerabend and Lakatos that competition among theories (or research programs) stimulates the growth of science but I admit the existence of evolutionary periods in which one theory dominates. Besides, I reject Feyerabend's anarchistic attitude and support Popper's and Lakatos' search for the logic of science (we shall return to this point in Ch. 9).

7.6. THE CONCEPT OF REVOLUTION AND ANTI-CUMULATIVE CHANGES

The concept of revolution needs precisioning. Which changes in science should be called 'revolutions'? There are different approaches to this question.

One approach consists in strong demands submitted to revolution, e.g. demand for the change of basic laws of science or even of its basic methodological principles. An extreme view is represented by I.A. Akchurin who writes that a revolution occurs only when all basic methodological problems (he distinguishes four classes of these problems) obtain new solutions (Akchurin, 1973, p. 222). If we used such a narrow concept, revolutions would occur very seldom.

Some authors connect revolutions with meaning-variance. S. Amsterdamski distinguishes three concepts: *evolution, local revolution* and *global revolution.* In the first case the meaning of concepts remains unchanged, in the second case some concepts are semantically reinterpreted, in the third case general concepts are reinterpreted and in consequence of it the whole world-image changes (Amsterdamski, 1973, Ch. 7, 8). It is, however, not clear when there is a semantical reinterpretation and when there is not, which concepts are general, etc.

Another approach is more liberal. It connects revolution with some changes in the set of statements and concepts accepted in science.

An example is Roman Suszko's conception based on the diachronic logic elaborated by him in the 50's. We shall present the main ideas of this conception.

Suszko considers knowledge to be an epistemological relation (E-opposition) between the subject S and the object M, or between the language L (as the main element of S) and its model M. The model is composed of the universe U and its characteristics. A set of sentences T asserted by the subject is a component of S. A set of axioms A is also distinguished ($A \subset T$). The E-opposition appears in the following form: $((L,A,T)\ M)$. Its transformation in the course of the development of science is designated as follows: $((L,A,T),M)/((L^x,A^x,T^x)M^x)$. Suszko considers three kinds of transformations: (1) evolutionary changes, (2) weak revolutions, (3) strong revolutions.

(1) In the case of *evolution* M and L do not change. Only T changes: some new theses are formulated, some old ones (which turned out to be false) are dropped. A may change or not. The transformation has the following form: $((L,A,T),M)/((L,A^x,T^x),M)$.

(2) In the course of a *revolution* M and L change too. In the case of a *weak revolution* the model expands (M is a submodel of M^x). The language expands also: the universe U is unchanged but new concepts are introduced (they catch new properties and relations).

(3) In the case of a *strong* revolution the universe is growing ($U \subset U^x$). Among the new constants of L^x there is a constant ϕ which denotes the old universe: $d(\phi) = U$. A relativization to this constant is taken into account: the unlimited general quantifier $\bigwedge\limits_{x}$ is replaced by the limited one $\bigwedge\limits_{x}$ (if $\phi(x)$ then . . .) (Suszko, 1957, 1968).

Critics noticed that Suszko's conception does not take into account the elimination of some concepts (e.g. of phlogiston) from science (Giedymin, 1968). This objection is perhaps not very important because in contemporary science such eliminations, probably, do not occur. Another objection is more serious: Suszko's weak revolution falls entirely within the set of changes usually called evolutionary (Kuhn's normal science, Lakatos' development of the research program) (cf. Amsterdamski, 1973, Ch. 7). Only Suszko's strong revolution deserves the name of revolution. The relativization to ϕ means here a limitation of the old laws to a narrower range (e.g., to $v \ll c$ in STR). However, as we know, the situation is not so simple: old laws turn out to be not exactly but only approximately true even in this narrower range. Apart from it, in the course of revolution science passes to a new idealization, to an essentially new model —Suszko's model does not reveal this complex process.

Some authors hold the concept of revolution to be unclear, hence they try to replace it by other concepts.

Elżbieta Pietruska distinguishes cumulative and anti-cumulative changes in science. She considers the vocabulary V of the language used in science, the set T of sentences held to be true and the extension E of concepts used in them.[9] Everywhere she puts the index '1' for an earlier period and the index '2' for a later one.

There are three kinds of *cumulative* changes:

(1.1) The set of admitted sentences is enriched (new laws are discovered): $T_1 \subset T_2$.

(2.1) The vocabulary is enriched (new concepts are introduced): $V_1 \subset V_2$.

(3.1) The extension of a concept is widened (new objects of the given kind are discovered): $E_1 \subset E_2$.

We may add that new sentences are introduced into science not only in case (1.1) but also in case (2.1) (laws containing new concepts) and in case (3.1) (sentences ascribing the properties of the given kind to new objects).

There are three analogous kinds of *anti-cumulative* changes:

(1.2) The set of accepted sentences is narrowed (some of them turn out to be false: $T_2 \subset T_1$) or, what happens more often, it is widened and shortened at the same time, i.e. there is an intersection of both sets of sentences: $T_2 - T_1 \neq \varnothing, T_1 - T_2 \neq \varnothing, T_1 \cap T_2 \neq \varnothing$.

(2.2) The vocabulary is shortened (some terms are dropped: $V_2 \subset V_1$) or, what happens more often, there is an intersection of both vocabularies: $V_1 - V_2 \neq \varnothing, V_2 - V_1 \neq \varnothing, V_1 \cap V_2 \neq \varnothing$.

(3.2) The extension of a concept is narrowed (some objects turn out to be incorrectly subsumed under this concept: $E_2 \subset E_1$) or, what happens sometimes, there is an intersection of both extensions: $E_2 - E_1 \neq \varnothing$, $E_1 - E_2 \neq \varnothing, E_1 \cap E_2 \neq \varnothing$.

It is again clear that some sentences are eliminated from science not only in case (1.2) but also in case (2.2) (theses containing abandoned terms) and in case (3.2) (sentences wrongly ascribing properties of the given kind to some objects).

Pietruska observes that all six kinds of changes occur in science. The first three are quite coherent with the simple cumulative model of science. The last three are not. They comply with the revolutionary model of science. However, when may we speak about a revolution? Probably, when anti-cumulative changes are intense enough. But what measure of their intensity should we choose? There is no answer to this question. Hence, the Kuhnian concept of scientific revolution is notoriously not sharp. Therefore, it is better to abandon this concept and to speak only about cumulative and anti-cumulative changes in the history of science (Pietruska-Madej, 1977).

I think that Pietruska's concepts are very useful for analysis of the history of science but her resignation from the concept of revolution is too rash. The distinction between evolution and revolution in science can be analyzed in these concepts.

Cumulative changes of all three kinds prevail in science. Probably, in long periods they alone occur. Perhaps also anti-cumulative changes of the kind (3.2) (e.g. a discovery that a substance does not belong to a class to which it was assigned previously by chemists), though they are incompatible with simple cumulativism. We call these periods evolutionary.

Consider now the two other kinds of anti-cumulative changes. A change of the kind (2.2), i.e. elimination of a concept, is typical for the transition through the threshold of maturity: elimination of Ptolomy's epicycles, of Aristotle's natural motions, of Stahl's phlogiston, etc. (to be sure, sometimes this elimination is not immediate: Copernicus used the concept of epicycle, later on Kepler abandoned it). In mature science it is an exceptional event: twentieth century physics abandoned the concept of aether (I don't know any other example). I think that elimination of a concept from science is always connected with a revolution.

Changes of the kind (1.2) are more usual. Each discovery that a law accepted in science turns out to be false (in its previous formulation) is a revolution. As we know, in mature science there is always a CR between T_1 and T_2: some laws of T_1 turn out to be false in a previously alleged domain, however they preserve their validity in a narrower domain with some approximation or preserve it exactly in a new ideal model. Hence, we may subsume these changes under the label (1.2) only with some qualifications.

Of course, when we say that elimination of a concept or refutation of a law is a revolution we mean concepts and laws commonly accepted in science. Rejection of concepts (e.g., Blodlot's N-rays) or laws (e.g., Lyssenko's theses) accepted only by some scientists but not by all the scientific community is not a revolution.

NOTES TO CHAPTER 7

[1] S. Toulmin illustrates this idea with the following example. Darwin's theory of evolution was a synthesis which went beyond two one-sided conceptions in geology and paleontology: Cuvier's catastrophism and Lyell's uniformitarianism (Toulmin, 1970).

[2] B. Kedrov even claims that Kuhn has brought out, in different terminology, ideas elaborated long ago by Marxism. Kuhn's book contains, therefore, no basically new ideas; nevertheless, it is valuable in a philosophical respect because Kuhn confirmed, perhaps without intending to, the Marxist conception of the sequence of evolutionary and revolutionary periods in development with a great deal of material from the history of science (Kedrov, 1969b, p. 28). I cannot agree with Kedrov: Kuhn has provided a new and deep, though controversial in many points, analysis of the structure of scientific revolution; this analysis goes far beyond the general philosophical statements contained in previous Marxist literature.

[3] N. Koertge points out a correspondence between Stahl's phlogiston theory and Lavoisier's oxygen theory: each chemical equation in the former has a corresponded equation in the latter (Koertge, 1969, Section 6). However, the term 'correspondence' is used here in the meaning of a set theory. The meaning of this term in CP is much narrower.

[4] S. Amsterdamski writes that in the case of the differential form of Ohm's law there was no revolution because the meaning of the concepts was not changed and in the case of STR there was a revolution because the meaning of basic concepts was changed (Amsterdamski, 1974, Ch. 7). I do not know why in one case the meaning is changed and in another it is not.

[5] B. Kedrov in a recent paper distinguishes two types of initial scientific revolutions: (1) the transition from appearance to essence (e.g., Copernicus in astronomy, Lavoisier in chemistry), (2) the revelation that a given sphere of Nature is not stable but changeable (e.g. Kant and Laplace in astronomy, Lamarck and Darwin in biology). Sometimes both revolutions were connected (R. Mayer revealed the essence of heat and its transformations in other processes) (Kedrov, 1974). The first type is identical with my concept of initial revolution. The second type has a different character because we deal here not with rejection of a theory but rather with creation of a theory of a new type. Nevertheless, I agree that it is a kind of scientific revolution as well (a radically new view on phenomena), and very important from the philosophic standpoint.

[6] L. Krüger distinguishes also two types of revolutions in almost the same way as I do: (1) refutation of a false theory, (2) correction of an insufficient but 'to a certain content true' theory (Krüger, 1973). He even gives partly the same examples. The similarity is striking, however we have come to this conception independently: Krüger presented it at the Bucharest congress in 1971, but I had no information about it when I wrote my Polish paper on this subject in 1973 (Krajewski, 1974a). There is, however, a difference between us in the interpretation of the second type of revolution. Krüger said simply that in this case the new theory is compatible with its predecessor. This thesis was attacked in the discussion. Krüger mentioned limit transition but did not speak about CR; he did not see that between T_1 and T_2 there is both contradiction and compatibility, i.e. a correspondence in the sense described in the previous chapters of our book.

[7] As J. Watkins correctly observes, Kuhn is the author of one excellent book on the Copernican revolution and of another more famous book on scientific revolutions in general; hence, it is very strange that he dislikes scientific revolutions (Watkins, 1970).

[8] S. Novaković says that creation of a new theory cannot be sudden and therefore periods of revolution are not favorable to creation of theories. He admits that the acceptance of a theory alone can be a sudden process (Novaković, 1971, p. 193). I agree that the formation of a new theory is usually accomplished by many scientists' uphill work (Einstein's GTR is an exception) but this work is most intense in periods of crises and revolutions. On the contrary, acceptance of a theory is not sudden: usually after a revolution adversaries of the new theory persist for a long time.

[9] I change her designation to adopt it to the English language.

CHAPTER 8

RELATIVE AND ABSOLUTE TRUTH

8.1. RELATIVE TRUTH

The basic aim of science is, first of all, the truth. We shall now consider this concept more carefully.

We read in numerous books on the philosophy of science that a law-like statement, if true, is a law of science. However, this claim is too simple and naive. As many authors have noticed, science usually does not reach the truth. Its quantitative laws are probably always (or almost always) false in the rigorous classical sense. Nevertheless, they are not mistakenly accepted in science. They are more or less 'close' to the truth. To describe this relationship, many philosophers introduce a concept of relative truth, approximative truth, verisimilitude, etc. Such a concept seems to be inevitable if someone wants seriously to analyze real science.[1] Otherwise, we fall either into dogmatism, considering the last theory as true, or into scepticism, considering all scientific theories as equally false, or into relativism, eliminating the concept of truth as a relation to reality altogether.

The concept of a relative truth is one of the basic concepts of the epistemology of dialectical materialism. However, this concept is used in Marxist literature in different ways, which has been analyzed by some authors, mainly in Poland (Eilstein, 1963; Krajewski, 1963 and others). Often a statement is held to be relatively true in the sense that in some conditions it is true, in others it is false (the truth is 'concrete'). However, it is a trivial situation, from the logical point of view, and we may say that the statement is formulated imprecisely (it is 'elliptic'). When we bring out the conditions more explicitly, the statement will turn out to be either true or false. Sometimes the concept of relative truth is used in the meaning of a hypothesis, of a not absolutely certain statement. This case is also uninteresting from the logical point of view. We shall not use the expression 'relative truth' in the meanings mentioned above.

However, there is a very important meaning of this expression, used often in the classics of Marxism: relative truth as an approximation. A scientific law, e.g. a gas law, is relatively true because it is approximate; it contains 'grains of truth' and 'grains of falsehood', hence it may change, it is not final. In the course of the growth of knowledge the grains of truth become more numerous, science is approaching absolute truth, but this process is infinite, it has an 'asymptotic' nature (Engels, 1878, Part I, Ch. IX; Lenin, 1909, Ch. II, §5).

There were many discussions in U.S.S.R, Poland, G.D.R. – mainly in the 50's – on the relationship of the concept relative truth to two-valued logic. Some authors proposed the use of multi-valued logics, others denied any possibility whatsoever of agreement between formal logic and dialectics. Now the majority of Marxist philosophers (in any case of philosophers of science) sees no conflict between dialectic and formal logic and uses two-valued logic. However, the concept of relative truth is not yet explicated satisfactorily.

Relativity of truth is associated with the role of the subject which always influences the content of knowledge. The common sense image of the world is determined in a high degree by our receptors and our place in the cosmos. Science is 'geocentric' (cf. Engels, 1925), at least in its initial stages. In the course of science's progress the subjective components of knowledge decrease, the image becomes more and more objective. In this sense we may also speak about movement through relative truths towards absolute truth (cf. Rainko, 1967, 1971).

In non-Marxist literature the problem of the relationship of relative and absolute truth in science was scarcely noticed for a long time. True, many relativists claimed that truth is always relative, depending on the conditions or on the subject, but they abandoned the concept of absolute truth completely. Their adversaries defended objectivity of truth and two-valued logic, they claimed that complete (non-elliptic) statements are always either true or false (e.g. Twardowski, 1900). However, they did not notice the problem of approximate truth in science.

One of the exceptions was P. Duhem who wrote that a law of science is neither true, nor false, but approximate (Duhem, 1906, Part II, Ch. 5).[2] However, he did not expound on these concepts, nor did he speak about movement towards absolute (not approximate) truth.

In our time this problem is noticed more often. E. Nagel writes that no theory is a 'final truth' about the 'ultimate nature' of things, that even a false theory may be useful, that the succession of theories in a given branch of science is a series of progressively better approximations to the unattainable but valid ideal of a finally true theory (Nagel, 1961, pp. 139, 143-44). M. Bunge writes that every law formula is an approximation to reality and that a sequence of such formulae tends to an unknown and unattainable ideal limit of perfect adequacy to the objective pattern (Bunge, 1967, Vol. I, p. 346). Nagel's 'final truth' and Bunge's 'perfect adequacy' are synonyms of Engels' 'absolute truth', their tending to the unattainable ideal limit is very close to Engels' asymptotic approaching the absolute truth.[3]

M. Bunge sketched a formalized theory of the *partial* (relative) truth. He considers the truth-value V of a proposition p in a system (theory) S as a function $V(p/S)$ which takes its values in the interval $(-1, 1)$; for a completely true proposition $V = 1$, for a completely false one $V = -1$, when $V > 0$, p is interpreted as 'partially true', when $V \approx 1$ as 'approximately true', etc. (Bunge, 1963, Ch. 8). However, it is not clear how we should measure V, e.g. in which case $V = 0$, etc. Bunge points out that the truth-value is not equal to the probability. Nevertheless, in some cases he treats the truth-value as a degree of confirmation, i.e. as a logical probability: for non-metrical generalizations the empirical truth-value depends upon the ratio of confirming cases in the sequence of tests. It seems that the truth-content (I prefer this expression to 'truth-value') has nothing to do with probability or confirmation. It is a measure of approximation (see below 8.3).

K. Popper introduced the concept of *verisimilitude*, opposing it to the concept of probability used by confirmationists. He defines versimilitude as a difference between the measure of truth-content and the measure of falsity-content, the truth-content of a statement being a class of its true consequences which are not tautological etc. (Popper, 1963, Ch. 10; 1972, Ch. 2). Popper's ideas are very important. He rightly stresses the fact that truth-content has nothing to do with logical probability (however, the last concept is, of course, very useful in other cases). But his definitions scarcely hold water. They have been criticized by other British philosophers of science.[4] Some of them even claim that the comparison of two theories with respect to their verisimilitude is not possible at all (Miller, 1975).

H. Putnam speaks about approximate truth of a theory held 'not exactly but with a certain degree of error' (Putnam, 1965, p. 206). His idea is very close to mine but he does not develop it.

Recently in Poland, Ryszard Wójcicki introduced a concept of approximate truth in a formal way, using many additional concepts (approximative model, approximative range of a theory, approximative structure, its idealization, its conservative precisioning, etc.) (Wójcicki, 1974, Ch. IV, §2). His conception is interesting but rather complicated and we shall not discuss it here.

Before presenting my view on relative truth I will consider the problem whether there are absolute truths in science.

8.2. ABSOLUTE TRUTHS IN SCIENCE

Absolute truth is, by definition, a final unchangeable truth, i.e. a truth in the rigorous classical sense. Some authors say that there are no absolute truths in science at all. However, they are wrong.

Engels wrote that there are absolute truths but they are platitudes (*Plattheiten*). He gave examples: two times two is four, the sum of angles of a triangle is equal to two right angles, Paris is in France, Napoleon died on 5 May 1821, people die without food (Engels, 1878). Lenin mentioned these examples criticizing the Russian relativistic-oriented Marxists who held any admittance of absolute truth to be dogmatism incompatible with the dialectic (Lenin, 1909).

Among Engels' examples are different kinds of statements. The first two mentioned are mathematical statements. We consider them now to be analytical, in definite axiomatic systems of course. The need for this qualification is striking with respect to the second example: we know that triangles have the mentioned feature only in the Euclidean system of geometry (Engels did not know about non-Euclidean geometries, knowledge of them was not yet widespread then).

The other three statements are synthetical. However, they are of a different nature. We shall distinguish three kinds of synthetical statements which can be absolutely true.

(1) *Qualitative facts.* We class different kinds of facts here: facts-events (e.g., Napoleon died), facts-states of affairs (e.g., Paris is in France), facts-relations (e.g., the Sun is bigger than the Earth). Accidental generalizations (e.g. all people in this room are adult) belong to this class too. All examples in parentheses are qualitative facts. They are either absolutely true, or absolutely false, they cannot be relatively true.

We speak about *qualitative* facts because there are also *quantitative* facts, e.g., Paris has five million inhabitants, the distance between Sun and Earth is 150 million kilometers. These statements are approximate, hence they are relatively true.

(2) *Existential* statements. Examples: there are metals heavier than water, there are mammals living in the sea. In general:

(8.2.1) $\bigvee_x [a(x) \wedge b(x)]$.

(3) *Qualitative laws* (qualitative strict general statements). Examples: all metals are good conductors of electric current, all gases can be liquefied, all living beings need food. In general:

(8.2.2) $\bigwedge_x [a(x) \Rightarrow b(x)]$.

The statements of all three above-mentioned kinds are either absolutely true, or absolutely false. We usually suppose that the statements formulated by us are absolutely true. Of course, this is not always the case – we are wrong sometimes. We often say that a sentence belonging to one of these kinds is probable, more probable than another one, etc. We are sometimes not sure about the facts when the evidence (witnesses, documents) is not definite. For example, we may treat as more or less probable the fact-statements: Caesar was in Great Britain, Oswald was the single killer of J.F. Kennedy, there are living beings on other planets, etc. However, as Popper correctly observes, the probability of a statement has nothing to do with its verisimilitude (relative truth). The statements about Caesar, Oswald and living beings cannot be relatively true, they are either absolutely true, or absolutely false. Only *we* are in doubt. In the case of relative (approximate) truth the situation is usually opposite: we are sure that an approximative law is not absolutely true (e.g. gas laws are not exactly true with respect to real gases).

Notice also that Engels was not right when he wrote that absolute truths are platitudes. Many synthetic statements of all three kinds cited above (the Sun is bigger than the Earth, there are metals heavier than water, there are mammals living in the sea, all gases can be liquefied) are not trivial; in any case they were not trivial when they were formulated for the first time. Many analytical statements of mathematics are not trivial either.

8.3. TRUTH-CONTENT AND APPROXIMATE TRUTH

Although qualitative statements are often not trivial, it is the quantitative ones which play the crucial role in modern science. Quantitative fact-statements are rarely absolutely true, and quantitative laws probably never so. In other words the main theses of science are only relatively true. *A fortiori*, scientific theories, containing quantitative laws, are never absolutely true. Therefore, the precisioning of the concept of relative truth is one of the main tasks of the philosophy of science. Unfortunately, this task is rarely taken up. I will propose a sketch for such a precisioning.

Many logicians, reluctant about the idea of the relativization of truth, ask how we can distinguish a relative truth statement from a false one. This problem is indeed not easy to solve.

In the case of quantitative fact-statements we can, it seems, solve this problem only conventionally. Probably, everyone will agree that the statement about the distance between Earth and Sun being 150 mln km is relatively (approximately) true and the statement about this distance being 50 mln km is false. In the first case the relative error is less than 0.5%, in the second one more than 50%. But where is the limit? The delimitation is conventional and in different cases may be different. Sometimes a precision of 10% is enough, sometimes we need 1%, 0.1%, etc.

Of course, we must further differentiate statements admitted as relatively true. We shall introduce a concept of truth-content but in a different way from the Popperian one. The truth-content (TrC) of a quantitative *fact-statement* grows with improvement of the approximation. When we designate the relative error made by a fact-statement F by E and the truth-content of F by $\text{TrC}(F)$ we may define it as follows:

(8.3.1) $\text{TrC}(F) = 1 - E(F)$.

We never know E exactly. However, we know the possible maximal error when a given measuring method is used. We may interpret E as the maximal relative error:

(8.3.2) $E = \dfrac{\Delta a}{a_1}$,

where a_1 is a result of measuring and Δa the maximal absolute error (the exactness of the measuring method).

In the case of a quantitative law the procedure is a little more complex. Suppose a law L provides a functional dependence $F(A,B) = 0$ between parameters A and B. We can predict a value b_1 of B by means of L when the value a_1 of A is given by the measuring. Since L is approximate (with respect to reality) the prediction is also approximate, even when we assume that the value a_1 is exactly known. To be sure, the approximation is different for different values of parameters. Therefore, we shall apply the concept of degree of inadequateness (DI) proposed by W. Patryas (see 6.5). However, it must be relativized to a parameter.

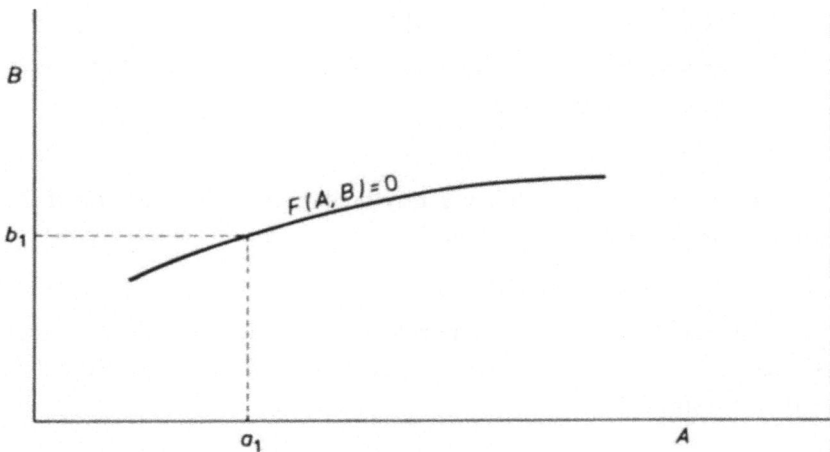

The DI of a law L, with respect to a parameter B contained in it, is equal to the Supremum of relative errors made by the use of L in prediction of various values b_i of B. If we designate the DI of L with respect to B by $DI_B(L)$ and the relative error made in the prediction of a value b_i of B using L by $E_{b_i}(L)$, we may write:

(8.3.3) $DI_B(L) = \text{Sup} \, [E_{b_i}(L)]$.

The truth-content of L must be, of course, also relativized to a parameter. If we designate by $TrC_A(L)$ the truth-content of L with respect to A, we may define it as follows:

(8.3.4) $TrC_A(L) = 1 - DI_A(L)$.

We may give an absolute definition of a truth-content of law $TrC(L)$ now. It is a minimum of all its relative truth-contents with respect to different parameters. If we designate by j any of the parameters contained in L, the definition is as follows:

(8.3.5) $\text{TrC}(L) = \text{Min}_j [\text{TrC}_j(L)]$.

Of course, TC has a minimum value when DI has a maximum value. Therefore, we may also write:

(8.3.5') $\text{TrC}(L) = 1 - \text{Max}_j [\text{DI}_j(L)]$

or in a more developed form:

(8.3.5") $\text{TrC}(L) = 1 - \text{Max}_j \{\text{Sup}_i [E_{ji}(L)]\}$.

We pass from laws to theories now. A theory T may be held to be a conjunction of laws contained in it:

(8.3.6) $T = L_1 \wedge L_2 \wedge \ldots \wedge L_n$.

The truth-content of a theory T may be defined as the minimum of the truth-content of all laws L_i contained in it:

(8.3.7) $\text{TrC}(L) = \text{Min}_i \text{TrC}(L_i)$.

Using traditional Marxist terminology, one may say that the truth-content of a theory is the amount of absolute truth (or of the 'grains' of absolute truth) in it.

8.4. THE TRUTH OF IDEALIZATIONAL LAWS AND OF THEIR FACTUALIZATIONS

All previous considerations concerned *factual* laws and their relationship to reality. Now we must consider the problem of the truth of *idealizational* laws.

We introduce the concept of Model Truth (MTr), as adequacy to a model, in opposition to classical truth (Tr) as adequacy to reality. As known, many philosophers and logicians use the concept of model truth (after Tarski) but they usually do not use the concept of classical truth: they consider only models and not objective reality; they are not concerned with the problem of testification of the truth by experiments (which can be done, of course, only with real objects). Other philosophers use only the concept of classical truth and its testification; they neglect modern science which usually has to do with ideal models. Hence, none of them considers the relationship between MTr and Tr. And the analysis of this relationship seems to be necessary. We can give only some general ideas on this analysis.

The set of all (factual and idealizing) assumptions of an idealizational law (the antecedent of (2.5.2)) forms an ideal model. We may say that an idealizational law is absolutely true in its appropriate model. However, we never know whether the model we have constructed is really *appropriate*. The construction of an appropriate model is an ideal case which occurs seldom if at all.

Consider the classical examples. The laws of CM were formulated for an inertial system and for a long time it seemed that they were absolutely true in this system. However, STR, GTR, and QM revealed additional idealizing assumptions which must be made in order to ascertain the truth of the law of CM, i.e. their absolute MTr. We are not sure whether these assumptions suffice.

We may hope that the gas laws are absolutely true in the model which is called ideal gas, but we are not sure whether this is actually so. The possibility that further investigations will lead to the necessity of enriching the set of idealizing assumptions now made by physicists is never excluded.

The absolute *MTr* of an idealizational law is possible, at least in an ideal case. Its absolute *Tr* is impossible.

In order to compare an idealizational law to reality, we usually factualize it. If we could take into account all side factors, the final factualization would give us the *absolute Tr*. However, the number of side factors is probably infinite in each real process, hence we cannot obtain an absolutely true factual law. The final factualization is always approximative and gives only a *relative Tr*.

An idealizational law may be also directly compared to reality. As we noticed in (2.7), within some range of conditions, namely when side factors are small, an idealizational law gives a sufficiently good approximation to reality, hence it is relatively true in a classical sense. However, its subsequent factualizations give better and better approximations, i.e. they have more and more TrC. When we designate an idealizational law by L_I and its subsequent factualizations by $L_I', L_I'', \ldots, L_F$ (see 2.5) we may write:

$$(8.4.1) \quad \mathrm{TrC}(L_I) < \mathrm{TrC}(L_I') < \mathrm{TrC}(L_I'') < \ldots < \mathrm{TrC}(L_F).$$

Of course, we may construct very different ideal models and formulate different idealizational laws. Not all models are satisfactory, not all alleged idealizational laws may be held to be relatively true. Is it possible to distinguish a satisfactory model from an unsatisfactory one, a genuine (relatively true) idealizational law from a false one, i.e. from an idealizational-law-like-statement which is not a genuine law? I believe that in principle this is possible by means of the concept of essence, of main and side factors.

8.5. RELATIVE TRUTH AND ESSENCE

When we admit the existence of main (essential) and side factors for each phenomenon, we may give the following definitions. A genuine idealizational law correctly grasps the main factor (the essence) of the given phenomenon. It is relatively true (in the classical sense). When an idealizational-law-like-statement grasps wrongly the main factor, i.e. mistakes a side factor for a main one, it is false. Revealing it, we must reject this statement (the alleged law) from science.

We return to our examples of initial scientific revolutions. Ptolomy's theory (GC) could predict the movement of planets: however, it was false because it grasped wrongly the essence of our planet system. Copernicus' theory (HC) was relatively true because it grasped this essence correctly: the revolution of planets around the Sun.[5] Copernicus' theory predicted the movement of planets with some approximation, not better than the last versions of GC, hence HC still had a high DI and a not very high TrC. Kepler's theory (KL) gave a much better approximation (ellipsis instead of circle); its DI was lower, its TrC higher than that of Copernicus' HC. Newton's theory (CM), applied to the solar system, predicted the movement of the planets better than KL because it took into account such side factors as perturbations caused by other planets; its TrC is higher than that of KL. Finally Einstein's GTR predicts the (Mercury's) perihelion movement not predicted by CM, hence GTR has the highest TrC. We may write:

$$(8.5.1) \quad TrC(Hc) < TrC(KL) < TrC(CM) < TrC(GTR).$$

Probably GTR is also not absolutely true. In a trivial sense this is obvious. It disregards electromagnetic and other forces which also affect planet movement although their influence is so weak that it cannot be measured.

The same may be said about the other examples of initial revolutions. Aristotle's mechanics is false because it grasps wrongly the essence of movement, CM is relatively true because it grasps this essence correctly. Stahl's phlogiston theory is false because it grasps wrongly the essence of combustion, Lavoisier's theory is relatively true because it grasps this essence correctly. Some qualitative laws contained in the theories of mature science are even absolutely true. E.g., Lavoisier's thesis that combustion is a synthesis is, probably, absolutely true. However, his law of conservation of mass in chemical reactions — a quantitative law — is only relatively true, when we consider it as Lavoisier did. He could not, of course, take the relativistic mass connected with energy into account.

When we have two essentially different models of a phenomenon we may expect that only one of them grasps correctly the essence of this phenomenon and enables us to formulate a relatively true idealizational law. Another model leads to a false idealizational-law-like-statement. Of course, in practice the decision is not easy; it may come after a long time. Each model must be thoroughly elaborated, idealizational laws based on different models, after their factualizations, must be compared with each other by means of experience. Eventually, experience will enable the decision. It is, as we see, the procedure recommended by the pluralistic idealizational empiricism described in 6.5.

8.6 TOWARDS THE ABSOLUTE TRUTH

We must distinguish two kinds of absolute truth (cf. Schaff, 1951). The statements considered in 8.2. belong to the first kind. They may be called *partial absolute truth*. As we have seen, they do not play a crucial role in science. They describe some qualitative features of things.

However, when we speak about absolute truth we mean in most cases another kind of knowledge: a theory which gives a complete description of reality, or at least, of a domain of reality. Such a theory is final, unchangeable, and hence absolutely true. It may be called *total absolute truth*. In contradistinction to the first kind, absolute truth of the second kind is unattainable in practice. It is only an ideal limit at which science aims.

In Marxist literature the second kind is usually taken into account. The classics of Marxism meant this when they wrote that absolute knowledge is reached in an infinite, asymptotic process (Engels, 1925), in an infinite series of generation (Engels, 1878), that absolute truth arises 'from the sum' of relative truths (Lenin, 1909), etc.

In other words, science is approaching in infinite progress the ideal of absolute truth. How is this progress made? Why is it infinite? I think we may give at least three reasons for this infinity.

First, science uses better and better measuring instruments, and discovers more and more precise laws (in our terminology: laws with lower and lower DI). The improvement of measuring methods has no limit, which is sufficient for the claim that progress is infinite. But this aspect is not very interesting from the philosophical point of view.

Second, the world is infinitely complicated, even 'at the surface'. Everything has, probably, infinitely many properties (e.g., dispositions to act on various things in different conditions); we discover them in turn. Every real process is infinitely complex because infinitely many side factors intervene; we take them into account one by one. This is sufficient to hold the world to be 'inexhaustible' in the cognitive respect.

Third – and this is the most important point though also the most controversial – the world is, probably, infinitely complex 'in depth'. Structure of matter consists of different levels; we reach them gradually. It is probable that the number of these levels is infinite. Were it so, we should go on forever discovering new fundamental elements of matter, its new basic laws. The number of scientific revolutions would be infinite.[6]

Lenin's famous expression that the 'electron is as inexhaustible as the atom' (Lenin, 1909, Ch. V, §2) is probably to be understood in this sense (however, it may simply mean that the number of properties of the electron is infinite). At some other place Lenin associated directly the infinity of scientific progress with the infinite complexity of essence; he wrote that human thought goes deeper, from appearance to essence, from essence of the 'first order' to essence of the 'second order', etc. without end (Lenin, 1933).

It is noteworthy that K. Popper came to an analogous conclusion. He writes that there is no ultimate essence of the world but that science seeks 'to probe deeper and deeper into the structure of our world', towards its 'more and more essential' properties (Popper, 1972, p. 196). The similarity is striking though Popper did not mention Lenin and perhaps did not know this passage.

When we take into account the idealizational nature of the basic laws of science, we may present the main track of the progress of knowledge as follows.

Scientific revolutions usually consist of revealing new idealizing assumptions of old theories, and creating new, more factual, theories, which are in CR with the old ones. We approach absolute truth along a correspondence sequence $T_1, T_2, T_3 \ldots$ (see 4.8). In such a sequence the truth content gradually increases:

$$(8.6.1) \quad \mathrm{TrC}(T_1) < \mathrm{TrC}(T_2) < \mathrm{TrC}(T_3) < \ldots .$$

At the same time we have new formulations of the initial theory, due to gradual revelation of its appropriate idealizing assumptions: $T_1, T_1', T_1'', \ldots$ This sequence also approaches absolute truth although in a different sense. We are dealing here with a model truth. When we introduce a concept of model truth-content MTrC, analogous to the concept of classical truth-content, we may write:

$$(8.6.2) \quad \mathrm{MTrC}(T_1) < \mathrm{MTrC}(T_1') < \mathrm{MTrC}(T_1'') < \ldots .$$

In other words, we approach final absolute truth on two planes: a *factual* and an *idealizational*. On the first plane we have increasingly more exact factual laws, we approach *absolute classical truth* in an infinite process. In the second plane we have more adequate formulations of idealizational laws, we approach *absolute model truth* in an infinite process. Both processes are closely connected and neither of them is possible without the other.

We may notice that simultaneous use of the concept of classical truth and the concept of model truth is necessary in order to obtain a real picture of science's progress.

NOTES TO CHAPTER 8

[1] The same idea was recently expressed by K. Popper in one of his last papers: '. . . we have reason to conjecture that Einstein's theory of gravity is *not true*, but that it is a *better approximation to truth* than Newton's. To be able to say such things with a good conscience seems to me a major desideratum of the methodology of the empirical sciences' (Popper, 1972, p. 335).

[2] Lenin notices that in Duhem's 'but' the beginning of a fallacy is contained, an efface-ment of frontiers between a scientific theory approaching objective truth and a fantastic, conventional theory like religion or chess theory (Lenin, 1909, Ch. V, §8). However, the latter theories are not approximate and Duhem's 'but' may be interpreted as an opposition to the claim that scientific theories are absolutely true. Of course, Duhem was not a materialist but he was approaching in many cases – as Lenin himself noticed – dialectical materialism.

[3] M. Bunge notices at some point (without mentioning Engels) that the expression 'tend asymptotically to the complete truth' is incorrect for different reasons (Bunge, 1963, p. 124). He is right, but this expression is not used in a precise mathematical sense but applied as a metaphor which helps to grasp the general idea.

[4] A critical analysis of Popper's concept of verisimilitude is contained in papers by P. Tichý, J.H. Harris, and D. Miller in *The British Journal for the Philosophy of Science*, Vol. 25 (1974), No.2.

[5] Some contemporary authors hold the GC and HC systems to be equivalent, equally true. However, they are equivalent only in the kinematic aspect but not in other aspects (dynamic, energetic, cosmogonic).

[6] Of course, the problem whether Nature is indeed infinitely complex is controversial. Some physicists believe that science is close to the discovery of final truth and its development is approaching the end (Feynman, 1965). It seems, however, that this is an opinion of the minority. The opposite opinion is more probable and we may hope that a moment never comes when science finishes its development.

CHAPTER 9

INTERNAL AND EXTERNAL HISTORY OF SCIENCE

9.1. INTERNAL AND EXTERNAL FACTORS

The concepts of internal and external factors in science are used in different ways.

Some authors (usually physicists and logicians) regard the relations inside a theory or among different theories as internal and the relation between theory and experience as external (e.g. Strauss, 1970).

Some others (usually sociologists of science) regard all relations inside a scientific community as internal and the relations between this community and other social groups as external (e.g. Kuhn, 1971).

Most authors (especially philosophers but also historians of science) use intermediary concepts: they regard all *cognitive* factors and relations (both theoretical and empirical) as internal and other factors and relations – sociological, psychological, economic, ideological – as external (e.g., Lakatos, 1971).[1] We shall use only the latter concepts.[2]

The traditional approach to the history of science was *internalistic*: historians described discoveries and inventions, successive formulations of laws and theories. At best they tried to discover regularities in the growth of science and in the changes in its methods (instruments, ways of reasoning, application of mathematics, etc.). All these factors are internal.

However, some historians noticed the influence of other domains of human activity on science, i.e. the role of external factors. The turning point was the rise of Marxism. *Historical materialism* reveals the dependence of the different 'forms of the social consciousness' ('higher floors' of the superstructure) on economic life (basis) and on politics ('first floor' of the superstructure). Sciences, especially the social sciences (humanities), are influenced by economics and politics, often by means of other domains of social consciousness: philosophy, religion, etc.

On the other hand, *dialectical materialism* (created also by Marx and especially by Engels, although the term was first used explicitly by G. Plekhanov and popularized by V. Lenin) states that knowledge reflects objective reality, moves through relative truths towards absolute truth, etc. It stresses also the fact that knowledge of the laws of Nature is a necessary condition for dominating Nature. Material activity, practice (industry, experiment, etc.), is the basis for the rise of knowledge, its final aim and the criterion of its truth.

The great role ascribed to human material activity is the joining point of both parts of Marxist philosophy. Nevertheless, in the study of science they turn our attention in different directions: *dialectical materialism* towards internal factors and the absolute side of knowledge, *historical materialism* towards external factors and the relative side of knowledge. Epistemologic-ally-oriented Marxist philosophers usually stress the first side, sociologically- and historically-oriented ones the second.[3]

The influence of *external factors* on the development of science is un-doubted. All prominent contemporary historians of science see this, even those who are usually ranked among the 'immanentists'. A. Koyré asserts that science as a spiritual activity cannot be explained by material factors (as Marxists claim), nevertheless he admits the role of economic factors as stimuli for the growth of different branches of science; he stresses especially the dependence of scientific theories upon philosophy, upon the general fashion of thinking in a given period (Koyré, 1961, 1966). The dependence of science upon philosophy, upon the whole 'intellectual climate' of an epoch, is accentuated also by other distinguished historians of science (Butterfield, 1958; Crombie, 1959; Cohen, 1960; Rupert Hall, 1962; B.G. Kuznietsov, 1964).

Marxists and scholars influenced by Marxism often show the influence of social, especially economic, conditions on the development of science, namely the role of class differentiation in Antiquity, the role of the peculi-arity of different forms of social organization (e.g., in Europe and in China), the role of the rise of industry in the period of early capitalism, etc. (Farring-ton, 1949; Needham, 1953; Bernal, 1954; Zilsel, 1941). It is noticed that social conditioning of scientific thought may be either conscious, when a scientist directly responds to a social order, or unconscious, when his way of thinking is affected by social conditions (Lilley, 1953). This last phe-nomenon is especially investigated by the sociology of knowledge created by Karl Mannheim. He stressed (under the influence of Marx) the dependence of human mental activity (including science) upon class interest and has come (in contradistinction to Marx) to complete relativism, to a denial of the objectivity of knowledge (Mannheim, 1929).

At the same time *internal factors* play an essential role. Even art, religion, morality have their own 'logic', their internal regularities of development; according to the widespread terminology in Marxist literature, forms of social consciousness have a 'relative autonomy' in spite of their dependence upon the economic basis. There is continuity in the history of culture in spite of deep revolutions in it. All attempts to break up this continuity (futurism with its 'burning the Louvre', Russian 'proletcults' in the early Soviet period, the Chinese 'cultural revolution' in the 60's) are obscurant and doomed to fail. *A fortiori* it concerns science which has truth as its basic aim.

Nowadays, probably no-one denies the existence of both internal and external factors. Nevertheless, the problem of their relations raises controversies.

Lately the two approaches to the history of science are vividly discussed by philosophers and historians of science. We shall quote some opinions from the discussion in Minneapolis in 1969.

H. Feigl associates the internalistic approach with the context of justification, the externalistic one with the context of discovery. The former stimulates logico-methodological analyses, the latter historico-sociological ones. A good history of science must contain both of them (Feigl, 1970).

E. McMullin distinguishes between two concepts of science: S_1 a collection of propositions, S_2 a kind of activity. There are respectively two approaches to the history of science: the first one (HS_1) describes the results of scientific investigations, the theses and their evidences, the second one (HS_2) is an attempt to understand how S_1 has come to be. HS_1 focuses on the final versions of published works, HS_2 examines manuscripts, letters and different circumstances of the research. HS_1 is usually practiced by scientifically educated historians of science, HS_2 by the historically educated (McMullin, 1970).

Mary Hesse notices that internal HS considers science a history of rational thought about Nature, evolving according to its own inner logic, external HS considers science a social and cultural phenomenon inseparable from the whole social life. The former sees a close tie between science and philosophy, it stresses the role of epistemological criticism, while the latter loosens this tie (it is associated with a pragmatist or instrumentalist epistemology). The former searches for reasons admitted consciously by scientists in their work, the latter for causes (partial causes) which affect the development of science and usually remain unconscious. The two approaches are not incompatible. There are both reasons and causes in science. And probably no Marxist or Freudian would claim that every argument can be explained externally. However, in some cases a conflict arises, e.g., when someone claims that only external factors are decisive and the internal logic of science is a delusion. The author criticizes this view and defends the validity of internal HS (Hesse, 1970).

Analogous discussions take place among Marxists. Those who deal with historical materialism (sociology, history) prefer external HS, while those who deal mainly with dialectical materialism, with epistemology and philosophy of science, usually prefer internal HS.

Some points are accepted by both groups.

Economic and other practical needs exert influence on the rate of growth of different branches of science, on the problems taken on in the first line, etc. This is now commonly admitted. However, do practical needs determine the contents of the scientific theories? Some Marxist philosophers in Poland and other countries answer this question positively. They say that scientific theories depend to a great extent upon their practical aims and the social conditions in which they are elaborated. Hence, philosophy of science must be based mainly on sociology. In Poland certain philosophers (especially historians of philosophy) criticize the Poznań school for examining only internal relations in science and not their social functions, their relations to practical aims (e.g. Miś, 1974).

I cannot agree with this view. All theories must be evaluated from the point of view of truth, even in the social sciences, in which the content of theories is indeed partly determined by social factors, by interests of classes or other social groups. In the natural sciences the content of theories is scarcely influenced by class interest and other social factors. Social and psychological factors influence the choice of the subject of investigation, they can hamper or foster it, they can determine which hypotheses are examined at first — but they cannot determine the result. The results, the conclusions are determined in natural sciences by experience, by facts. And in the social sciences finally — too.

This view is probably obvious to scientists. It is also held by many philosophers, including Marxists. We shall provide three quotations from Soviet authors.

The first one is derived from a program paper published before the XIII Congress of the History of Science in Moscow: 'Practical needs give stimuli for the development (of science – W.K.) but it goes according to its own internal laws' (Mikulinskij and Markova, 1971, p. 110). The authors criticize both 'immanentist' and 'vulgar-sociologist' approaches to history of science. I agree with this attitude but I think that Koyré, called by the authors 'a leader of the immanentist stream', would agree with the quoted sentence too.[4]

Another Russian philosopher of science goes even further. He writes that many problems in science are created by practice but, of course, not all of them. 'The great problems of science . . . are due not to a "social order", they rise with a logical necessity from the course of development of the science itself, they have an immanent genesis' (Rodnyj, 1966, p. 25).

An Armenian author writes that different external factors – political, ideological, economic, organizational – may foster or hamper the growth of science, but the source of this growth is internal. Science surmounts, sooner or later, external obstacles, e.g. dogmatic ideological interventions. Its main internal stimulus is a 'drive to the ideal' (towards the future) connected with the Correspondence Principle (the link with the past) (Manasian, 1973).

Of course, different values direct scientific activity but truth is always the main one.

9.2. THE PROBLEM OF METHODOLOGICAL HISTORICISM

The idea of historicism plays a great role in Marxism and in some other contemporary philosophical schools. According to this idea, when we investigate a human phenomenon we must consider its genesis and development, its changes in the course of history, and we must search for the social conditions determining its shape in a given period. Science and its methods are human creations, therefore the historical approach is necessary here too. However, the interpretation of this approach raises many controversial issues.

It is undoubted that the methods of science change in the course of history. In different periods there are different methodological paradigms, to use the Kuhnian term. However, are these changes radical and irregular, as Kuhn and Feyerabend claim, or rather gradual and directed? Are they complete in the sense that no methodological norm is eternal, or are there any constant norms, common features in the methodology of science, since science comes to existence as such?

S. Amsterdamski discusses this issue in his last book. He does not agree
with the Kuhnian idea of a complete and radical variance of theoretical and
methodological paradigms in the course of scientific revolution. Though
some paradigms are changed, many others remain, which makes discussion
among adversaries in science possible; such a discussion does not stop during
a revolution, on the contrary, it is intensified. There are common methodological rules in every epoch, nevertheless, not in *all* epochs. There are no
'supra-historical' methodological rules, no eternal criteria for demarcation of
science from non-science (Amsterdamski, 1973).

I see in Amsterdamski's interpretation of historicism certain elements of
relativism, though, of course, not so strong as in Kuhn's case.[5]

If in remote epochs the criteria were quite different from ours, then the
theories from those epochs could not be compared to ours; it would be
impossible to claim that our theories are more scientific, closer to the truth,
etc.[6]

My view may be presented as follows. Methodological criteria change in
the course of history. Sometimes they are narrowed (e.g., the transition
from ancient and medieval science to the exact science of the seventeenth
century), sometimes they are widened (e.g. the surmounting of mechanistic
limitations and admission of statistical laws, non-central forces, etc. in the
nineteenth century). The evolution of methodological criteria is not yet
finished. Maybe it will go on forever. Nevertheless, it does not mean that
everything may change. There are most general criteria which are conditions
sine qua non for science. They form a 'hard core' of scientific methodology.
We shall consider two such criteria.

The first is the claim to *truth*. Mentioning this question, Amsterdamski
writes that the mere concept of truth changes. I do not think that these
changes are essential. The classic concept of truth as adequacy with reality
was abandoned by some philosophers but never by science. This concept
changes only in the sense that it is expressed more precisely than in the past.
A. Tarski has great merits here. His definition of truth is a precisioning of
Aristotle's (by using the concept of meta-language); there is continuity here,
not replacement. Of course, theories admitted in science are often incompatible with some facts; when these contradictions are not easily removable,
we speak about anomalies. However, scientists always hope to remove these
anomalies. In no epoch was incompatibility with facts accepted. Medieval
scientists wanted to 'solve the phenomena', the contemporary ones are more

rigorous in this respect.

The second general postulate is the consistency of scientific theories, the absence of logical contradictions inside them and among them. This postulate also is not always fulfilled. There are often contradictions between two theories, sometimes even inside one theory, but scientists treat such a situation, after they have noticed it, as alarming. Its removal is an urgent task.

In other words, scientists always want to and try to remove both external and internal contradictions (to use Strauss' expression) in science. These two postulates (of truth and of consistency) are valid in all epochs, they are 'supra-historical' (to use Amsterdamski's term). Probably, the 'hard core' of scientific methodology contains more than these two postulates.[7]

We see that historicism in methodology must not be treated in a one-sided way. The methods of science change but some of their features remain. The situation is analogous to the development of Nature and society. In spite of the evolution of Nature some basic features of matter, its fundamental laws, remain unchanged. In spite of the development of society some of its basic features, its general laws, also remain unchanged. We can see the unity of variableness and constancy everywhere. Relativism is as wrong as absolutism. One-sided historicism is as wrong as a-historicism.

9.3. INTERNAL HISTORY AS AN IDEALIZATION

Each logical scheme of the growth of science, each internal history of it is an idealization. It neglects external factors (sociological, ideological, psychological, etc.), i.e. it makes the idealizing assumption that they do not exist. This procedure is quite justified. Moreover, it is necessary when we want to develop the methodology and history of science as a genuine science.[8]

Some authors criticize the logical scheme of the methods and growth of science (e.g. schemes proposed by Popper, Lakatos, Poznań school, etc.) as inadequate. They point out that many facts from the real history of science are not explained by these schemes. These critics do not understand that these schemes are of an idealizational nature.

Of course, all idealizational theories must be confronted with experience. The history of science is not an advanced science, it has no mathematical shape, therefore, the 'normal' way of factualization and testification of factualized theories is impossible here. Nevertheless, different schemes of the growth of science must be evaluated. A scheme is better when it approximates real history more closely, i.e. explains more historical facts (cf. Lakatos, 1971).

Our model of the progress of science ought to be compatible in basic outline with the real history of advanced sciences. When it is not, it must be criticized. However, real history is never *completely* compatible with a model — numerous deviations are inevitable. In this case we should criticize rather the history.

This is not a paradox. Practice is never fully reasonable. It always deviates from rational schemes, e.g., explanations practiced in science often do not fulfill the Hempel-Oppenheim scheme of deductive explanation or even the scheme of probabilistic explanation. In these cases we must criticize the practice.[9]

There are three approaches to the philosophy of science.

The first may be called *crudely empirical*. It consists of bare descriptions of the methods used in sciences or of their factual history. The description of methods — as a matter of fact, only of methods used in the simplest cases in non-advanced sciences — and their inductive generalization was practiced by F. Bacon and J.St. Mill. We may call their attitude an *inductivist variant* of the crudely empirical approach to the philosophy of science.

Description of the factual history of science is a job for historians of science. It is all right when they do not claim that another approach to the history of science is impossible. However, some philosophers, like Kuhn and Feyerabend, claim that factual history with all its complexity and variety of methodological criteria prevents the construction of a rational scheme of the progress of science. We may call their attitude a *relativist variant* of the crudely empirical approach to the philosophy of science.[10]

Two levels must be distinguished: (1) the *epistemological* level — the approach to the methods of science, (2) the *meta-epistemological* level — the approach to the methods of the philosophy of science (epistemology, methodology).[11] Bacon and Mill were crude empiricists (inductivists) both on the epistemological and meta-epistemological levels. Kuhn and Feyerabend are not crude empiricists on the epistemological level but they are on the meta-epistemological level.

The second approach to the philosophy of science may be called *aprioristic*. It consists of the construction of logical schemes of an ideal scientific procedure without confronting them with real science. This method is used by many contemporary philosophers, both rationalists and empiricists in epistemology. For the former, e.g. for representatives of Husserlian phenomenology, it is quite natural: they are apriorists both on the epistemological and on the meta-epistemological levels. For the latter, e.g. for neopositivists, it is a little paradoxical. As is known, representatives of classical neopositivism were empiricists, even crude empiricists, in epistemology (with respect to empirical sciences). However, they constructed *a priori* schemes of a perfect empirical scientific procedure and did not want to compare them to real procedures used in science, in any case in advanced sciences. Hence, they were apriorists in meta-epistemology.

Classical neopositivism has passed away, nevertheless many logicians and philosophers of science continue to use its approach to methodology. They construct and analyze simple schemes of empirical procedures and do not confront them with the real sciences. As Jerzy Kmita correctly observes, they prefer to persuade scientists to do the confrontation of their theses with reality rather than to do it themselves (Kmita, 1972, p. 53).

We should notice, however, that some former members of the Vienna Circle do understand this now. Herbert Feigl even presents an honorific 'self-criticism' saying that he and some of his colleagues were 'sinners': proud of their empiricism, they 'made up' some phases of the history of science in a quite 'a priori' manner (Feigl, 1970).

The third approach to the philosophy of science may be called *idealizational-empirical*. It consists of the analysis of real science (first of all – of advanced sciences), of the creation of ideal models of its methods, of the confrontation of these models with real scientific procedures. It is the approach of all advanced sciences on a higher level. It is idealizational empiricism in meta-epistemology.

The crudely empirical approach in meta-epistemology is purely descriptive, the aprioristic approach is purely prescriptive, the idealizational-empirical approach is descriptive and prescriptive at the same time: it describes the methods used in science revealing their essence, grasping these methods in their pure form. Next it gives, though cautiously, some prescriptions. It shows deviations from the guidelines of science caused by different external (psychological, ideological, etc.) factors. It shows the future of less advanced sciences assuming that they will mature.

I am convinced that only the third approach is the right one. It was used in this book, in which a scheme of the progress of science is presented. This scheme is based on the Correspondence Principle and on the idea of the increasing Truth Content of theories combined with the idea of a competition among theories and a comparison of their Truth Content. This scheme is only sketched in this book. It must be further developed and made more precise for confrontation with the history of science.

I hope that this approach can be helpful to science. It will contribute an elaboration of an adequate rational image of the methods used in science, of the laws of its progress.

Philosophers cannot and must not evaluate scientific hypotheses and provide scientists with direct cues. They have done that sometimes in the past, usually with damage to science. The history of Marxist philosophy in the Stalin period gives many bad examples with respect to this. Nevertheless, philosophers can help science indirectly by means of analysis of its methods and development. However, only the scientists themselves can draw practical conclusions from this analysis.

NOTES TO CHAPTER 9

[1] Some authors rank 'transcendental conditions': lawfulness and knowability of the world, among external conditions of science (Böhme *et al.*, 1972).

[2] We may also speak about internal and external functions (or aims) of science. The internal (heuristic) function of science is the search for truth, for new knowledge (description, explanation, prediction, etc.). The external (practical) function of science is domination of Nature and, in general, fulfillment of social needs (cf. Nikitin, 1971, Ch. 1.2). Stress on one of these functions is usually associated with stress on one of two groups of factors determining the growth of science. Therefore, it is not fortuitous that, as we have mentioned in 7.5, Popper speaks of truth (the 'sky') and Kuhn of practice (the 'Earth'). The attention of the first is directed to internal factors, the attention of the second to external ones.

[3] There are Marxists who deny the difference between dialectical and historical materialism. They say that Marxist materialism is 'dialectical' and 'historical' at the same time (Bauman, 1956), the Marxist dialectic is a purely historical theory (Siemek, 1974), etc. They subordinate dialectical materialism to historical materialism, i.e. as a matter of fact, they liquidate epistemology (and *a fortiori*, the philosophy of science) as an autonomous part of philosophy.

[4] In a recent paper Mikulinskij speaks about two basic streams in the history of science: the externalist (historico-materialist) and the internalist (immanentist). The first, supported by the author, aims to find out the influence of social conditions on science, the second does not notice them. Nevertheless, the author points out that science itself formulates and resolves problems which are put forward by social needs, that each branch of science and the whole science has its own logic, its internal laws of development, etc. (Mikulinskij, 1974).

[5] The polemics with Kuhn was more sharp in Amsterdamski's postscript to the Polish translation of Kuhn's book (Amsterdamski, 1968). In the following years Amsterdamski took over some features of the Kuhnian approach.

[6] Many Marxist historians of philosophy in Poland go further in this direction than Amsterdamski does: they say that a given view cannot be considered at all outside of the epoch, i.e. the comparison of theories created in different epochs is meaningless. This view may be called 'extreme historicism' or 'one-sided historicism'.

[7] Some of my friends notice that there is only one postulate here, because the claim for truth entails the removal of contradictions. Nevertheless, I speak about two postulates in order to distinguish the adequacy to reality and the internal coherence. Both are important for empirical sciences; however, they raise different problems.

[8] E. Nikitin in his interesting book on explanation points out that every theory of the growth of science – like every theory of Nature – is not a mirror-reflection of reality (in this case of relations among theories in real history) but is constructed by means of a set of idealizations (Nikitin, 1970, 5.2.2).

[9] Not everybody understands this. E.g., S. Barker pointed out that the deductive scheme of explanation does not agree with many cases of real explanation made in science (Barker, 1961). Feyerabend rightly opposed him, claiming that in cases of incompatibility we should criticize the scientific practice and not the model. However, Feyerabend himself rejected the rational schemes of the growth of science as incompatible with real history. His inconsequence was correctly observed by Nikitin in the above-mentioned book (Nikitin, 1970, 5.2.2).

[10] E. McMullin says that Feyerabend's philosophy of science, despite its appearance, does not rest upon history, because he exaggerates some details of history, ignores others, etc. (McMullin, 1970). This is true; however, I speak about his general program, not about its realization.

[11] The distinction between epistemological and meta-epistemological approaches was made by Stanisław Rainko, though with a slightly different meaning (Rainko, 1972).

BIBLIOGRAPHY

The number after the name marks the year of the first original edition. The following abbreviations are applied:

LEP: The Library of Exact Philosophy
BSPS: Boston Studies in the Philosophy of Science
MSPS: Minnesota Studies in the Philosophy of Science
PSPSH: Poznań Studies in the Philosophy of the Sciences and the Humanities
PS: Philosophy of Science
BJPS: The British Journal for the Philosophy of Science
JP: The Journal of Philosophy
VF: Voprosy Filosofii (in Russian)
MF: Myśl Filozoficzna (in Polish, appeared till 1957)
SF: Studia Filozoficzne (in Polish)
CiŚ: Człowiek i Światopogląd (in Polish)
ZN: Zagadnienia Naukoznawstwa (in Polish)
KHNT: Kwartalnik Historii Nauki i Techniki (in Polish)

Achinstein, P.: 1964, 'On the Meaning of Scientific Terms', *JP*, Vol. LXI.
Achinstein, P.: 1971, *Law and Explanation*, Oxford.
Agassi, J.: 1963, *Towards an Historiography of Science*, The Hague.
Agassi, J.: 1966, 'Revolutions in Science, Occasional or Permanent?', *Organon*, No.3, Warszawa.
Ajdukiewicz, K.: 1965, *Pragmatic Logic*, Warszawa (English translation, Warszawa, 1971).
Ajdukiewicz, K. (ed.): 1965, *The Foundation of Statements and Decisions*, Warszawa.
Akchurin, I.A.: 1973, 'Some Regularities of the Development of Knowledge and Problems of Its Synthesis', in: *Synthesis of Contemporary Scientific Knowledge* (in Russian), Moscow
Akchurin, I.A. and Mamchur, E.A.: 1972, 'Logic of Discovery or Psychology of Research?' (Review of Lakatos and Musgrave, 1970) *VF* No.8.
Amsterdamski, S.: 1964, *Engels* (in Polish), Warszawa.
Amsterdamski, S.: 1968, Postscript to Polish translation of Kuhn, 1962, Warszawa.
Amsterdamski, S.: 1970, 'The Strife on the Problem of Progress in the History of Science', *KHNT*, No. 3.
Amsterdamski, S.: 1973, *Between Experience and Metaphysics*, Warszawa (English translation 1975, *BSPS*, Vol. XXXV).
Arseniev, A.S.: 1958, 'On the Correspondence Principle in the Contemporary Physics', *VF*, No. 4.
Augustynek, Z.: 1974, 'On the Correspondence Principle', in: Kmita, J. (ed.), *Methodological Implications of the Marxist Epistemology* (in Polish), Warszawa.
Barker, S.F.: 1961, 'The Role of Simplicity in Explanation', in: H. Feigl, G. Maxwell (eds.), *Current Issues in the Philosophy of Science*, New York.

Barr, W.F.: 1971, 'A Synthetical and Semantical Analysis of Idealization in Science', *PS*, Vol. 38, No. 2.

Barr, W.F.: 1974, 'A Pragmatic Analysis of Idealization in Physics', *PS*, Vol. 41, No. 1.

Bauman, Z.: 1956, 'Surmount the Disintegration of the Marxist Philosophy', *MF*, No. 6 (26).

Bazhenov, L., Morozov, K., and Slutskij, M.: 1966, *Philosophy of Natural Sciences*, Vol. I (in Russian), Moscow.

Berkson, W.: 1974, 'Some Practical Issues in the Recent Controversy on the Nature of Scientific Revolutions', *BSPS*, Vol. XIV.

Bernal, J.: 1954, *Science in History*, London.

Bohm, D.: 1951, *Quantum Theory*, Engl.Cliffs, N.J.

Böhme, G., van den Daele, W., and Krohn, W.: 1972, 'Alternativen in der Wissenschaft', *Zeitschrift für Sociologie*, Jg. 1, Heft 4.

Bohr, N.: 1922, *Drei Aufsätze über Spektren und Atombau*, Braunschweig.

Brittan, G.G. Jr.: 1970, 'Explanation and Reduction', *JP*, Vol. 67.

Bunge, M.: 1959, *Metascientific Queries*, Springfield, Ill.

Bunge, M.: 1962, *Intuition and Science*, Engl.Cliffs, N.J.

Bunge, M.: 1963, *The Myth of Simplicity*, Engl.Cliffs, N.J.

Bunge, M.: 1967, *Scientific Research*, Berlin.

Bunge, M.: 1970a, 'Theory Meets Experience', in: Kiefer and Munitz (eds.), *Contemporary Philosophical Thought*, Vol. 2, 'Mind, Science, History,' Albany.

Bunge, M.: 1970b, 'Problems Concerning Inter-Theory Relations', in: Weingartner and Zecha (eds.), *Induction, Physics, and Ethics*, Dordrecht.

Butterfield, H.: 1957, *The Origin of Modern Science 1300–1800*, London.

Cackowski, Z.: 1975, 'Remarks on the "Elements of the Marxist Methodology of Humanities" ', *SF*, No. 1 (110).

Causey, R.: 1972a, 'Uniform Microreduction', *Synthese*, Vol. 25.

Causey, R.: 1972b, 'Attribute-Identity in Microreduction', *JP*, Vol. LXIX.

Chmielecka, E.: 1977, *Context of Discovery and the Methodology of the Empirical Sciences*, in: Krajewski *et al.* (eds.), *Relations Between Theories in the Course of the Growth of Science*, Wroclaw.

Cohen, I.B.: 1960, *The Birth of a New Physics*, New York.

Colodny, G. (ed.): 1965, *Beyond the Edge of Certainty*, Engl.Cliffs, N.J.

Crombie, A.C.: 1959, *Medieval and Early Modern Science*, London.

Davydov, A.S.: 1963, *Quantum Mechanics* (in Russian), Moscow.

Duhem, P.: 1906, *Là théorie physique, son objet et sa structure*, Paris.

Eilstein, H.: 1958, 'Conception of Matter as Physical Being', in: Eilstein, H. (ed.), *The Material Unity of World* (in Polish), Warszawa.

Eilstein, H.: 1963, 'On the Conceptions of Relative Truth', *SF*, No. 2 (33).

Engelhardt, V.A.: 1970, 'Integratism – a Way from the Simple to the Complex in the Knowledge of Life Phenomena', *VF*, No. 11.

Engels, F.: 1878, *Herrn Eugen Dührings Umwälzung der Wissenschaft* (Anti-Dühring), Leipzig.

Engels, F.: 1925, (posthumously), *Dialektik der Natur*, Moscow.

Farrington, B.: 1953, *Greek Science*, London.

Feigl, H.: 1964, 'Philosophy of Science', in: *Philosophy*, Engl.Cliffs, N.J.

Feigl, H.: 1970, 'Beyond Peaceful Coexistence', *MSPS*, Vol. V.

Feyerabend, P.K.: 1962a, 'Explanation, Reduction, and Empiricism', *MSPS*, Vol. III.
Feyerabend, P.K.: 1962b, 'Problems of Microphysics', in: R. Colodny (ed.), *Frontiers of Science and Philosophy*, Pittsburgh.
Feyerabend, P.K.: 1965a, 'Problems of Empiricism', in: Colodny (ed.), *Beyond the Edge of Certainty*, Engl.Cliffs, N.J.
Feyerabend, P.K.: 1965b, 'Reply to Criticism', *BSPS*, Vol. II.
Feyerabend, P.K.: 1970a 'Consolation for the Specialist', in: Lakatos and Musgrave (eds.), *Criticism and Growth of Knowledge*, Cambridge University Press.
Feyerabend, P.K.: 1970b, 'Against Method: Outline of an Anarchistic Theory of Knowledge', *MSPS*, Vol. IV.
Feynman, R.P.: 1965, *The Character of Physical Law*, London.
Fine, A.I.: 1967, 'Consistency, Derivability, and Scientific Change', *JP*, Vol. LXIV.
Galileo, G.: 1642, *Dialogo sopra i due massimi sistemi del mondo, Tolemaico e Copernicano*, Firenze.
Garstka, W.: 1974, 'An Attempt of Formal Analysis of the Correspondence Principle', in: Krajewski *et al.* (eds.), *Correspondence Principle in Physics and the Development of Science*, Warszawa.
Giedymin, J.: 1968, 'Revolutionary Changes, Non-Translatability, and Crucial Experiments', in: Lakatos and Musgrave (eds.), *Problems in the Philosophy of Science*, London.
Giedymin, J.: 1973, 'Logical Comparability and Conceptual Disparity Between Newtonian and Relativistic Mechanics', *BSPS*, Vol. 24.
Gorski, D.P.: 1961, *Problems of Abstraction and the Formation of Concepts* (in Russian), Moscow.
Gorski, D.P.: 1973, 'Über abstrakte und idealisierende Objekte', in: Suppes *et al.* (eds.), *Logic, Methodology, and Philosophy of Science* (Proc. of the IVth Int. Congress for Logic, Methodology, and Philosophy of Science in Bucharest, 1971), Warszawa-Amsterdam.
Groenwald, H.J.: 1965, 'On Foundation of Elements of Physical Statements', in: Ajdukiewicz (ed.), *The Foundation of Statements and Decisions*, Warszawa.
Hanson, N.R.: 1958, *Patterns of Discovery*, Cambridge.
Hanson, N.R.: 1965, 'Newton's First Law: a Philosopher's Door into Natural Philosophy', in: Colodny (ed.), *Beyond the Edge of Certainty*, Engl.Cliffs, N.J.
Hempel, C.G.: 1965, *Aspects of Scientific Explanation*, New York.
Hempel, C.G.: 1966, *Philosophy of Natural Science*, Engl.Cliffs, N.J.
Hempel, C.G. and Oppenheim, P.: 1948, 'The Logic of Explanation', *PS*, Vol. 15.
Hesse, M.: 1970, 'Hermeticism and Historiography: An Apology for the Internal History of Science', *MSPS*, Vol. V.
Humphreys, W.C.: 1968, *Anomalies and Scientific Theories*, San Francisco.
Hutten, E.H.: 1956, *The Language of Modern Physics*, London.
Illarionov, S.V.: 1964, 'The Principle of Limitation in Physics and Its Connection with the Correspondence Principle', *VF*, No. 3.
Kagan, V.F.: 1944, *Lobatschevsky* (in Russian), Moscow.
Kałuszyńska, E.: 'On the Concept of Essence', *SF*, No. 5 (114).
Kard, P.T.: 1975, 'Principle of Non-Correspondence' (in Russian), *Acta et Commentationes Universitatis Tartuensis*, Tartu.
Karpinskaya, R.S.: 1971, *Philosophical Problems of Molecular Biology* (in Russian), Moscow.

Kedrov, B.M.: 1969a, *Three Aspects of Atomism*, Vol. I: *Gibbsian Paradox* (in Russian), Moscow.

Kedrov, B.M.: 1969b, *Lenin and the Revolution in Natural Science in the XXth Century* (in Russian), Moscow.

Kedrov, B.M.: 1974, 'Hegelian Dialectics in the Light of Scientific Revolutions', *VF*, No. 8.

Kedrov, B.M. and N.F. Ovchinnikov (eds.): 1974, *Problems of the History and the Methodology of Scientific Knowledge* (in Russian), Moscow.

Kemeny, J.G.: 1959, *A Philosopher's Look at Science*, Princeton, N.J.

Kemeny, J.G. and Oppenheim, P.: 1956, 'On Reduction', *Philosophical Studies*, Vol. VII, No. 1-2.

Kerszman, G.: 1958, 'On Engels' Conception of Forms of Movement', *SF*, No. 6 (9).

Kiefer, H.E. and Munitz, M.K. (eds.): 1970, *Contemporary Philosophical Thought*, Vol. 2, 'Mind, Science, History', Albany.

Kmita, J.: 1972, 'The Methodology of Science as a Humanistic Discipline', *SF*, No.1 (74).

Kmita, J. (ed.): 1974, *Methodological Implications of the Marxist Epistemology* (in Polish), Warszawa.

Koertge, N.: 1969, *A Study of Relations Between Scientific Theories: a Test of the General Correspondence Principle* (unpublished), London.

Koertge, N.: 1971, 'Inter-Theoretic Criticism and the Growth of Science', *BSPS*, Vol. VIII.

Koertge, N.: 1973, 'Theory Change in Science', in: Pearce and Maynard (eds.), *Conceptual Change*, Dordrecht.

Kołakowski, L.: 1962, *Notes on the Contemporary Counter-Reformation* (in Polish), Warszawa.

Korch, H.: 1972, *Die Wissenschaftliche Hypothese*, Berlin.

Kordig, C.R.: 1971, *The Justification of Scientific Change*, Dordrecht.

Koyré, A.: 1961, 'De l'influence des conceptions philosophiques sur l'évolution des théories scientifiques', in: *Etude d'histoire de la pensée philosophique*, Paris.

Koyré, A.: 1966, *Etudes galiléennes*, Paris.

Krajewski, W.: 1963, 'On the Concepts of Relative Truth', *SF*, No. 3-4 (34-35).

Krajewski, W.: 1968, 'Das Naturgesetz als notwendiger Zusammenhang', in: Kröber (ed.), *Das Gesetzesbegriff in der Philosophie und den Einzelwissenschaften*, Berlin.

Krajewski, W.: 1972a, 'An Outline of the Classification of the Laws of Science', *ZN*, Vol. VIII, No. 3.

Krajewski, W.: 1972b, 'Review of Nowak, L., 1971', *ZN*, Vol. VIII, No. 3.

Krajewski, W.: 1973a, *Engels on the Movement of Matter and Its Lawfulness* (in Polish), Warszawa.

Krajewski, W.: 1973b, 'Different Types of Theory-Reduction', in: *Proceedings of the XVth World Congress of Philosophy*, Sofia.

Krajewski, W.: 1973c, 'Correspondence Principle in Physics and the Development of Science', *KHNT*, Vol. XVIII.

Krajewski, W.: 1974a, 'Reduction, Idealization, Correspondence', in: Krajewski *et al.* (eds.), *Correspondence Principle in Physics and the Development of Science* (in Polish), Warszawa.

Krajewski, W.: 1974b, 'Mechanism and Reductionism', in: W. Krajewski (ed.), *From the History of Mechanism in the Physics and Chemistry* (in Polish), Wroclaw.

Krajewski, W.: 1975, 'Copernicus and Galileo versus Aristotle', *Colloquia Copernicana*, Vol. 4, Wrocław.

Krajewski, W. (ed.): 1969, *Concepts of Law of Science at the End of the XIXth Century* (in Polish), Wrocław.

Krajewski, W., Mejbaum, W., Such, J. (eds.): 1974, *Correspondence Principle in Physics and the Development of Science* (in Polish), Warszawa.

Krajewski, W., Pietruska-Madej, E., Zytkow, J. (eds.): 1977, *Relations Between Theories in the Course of the Growth of Science* (in Polish), Wrocław (in print).

Kröber, G. (ed.): 1968, *Das Gesetzesbegriff in der Philosophie und den Einzelwissenschaften*, Berlin.

Krüger, L.: 1973, 'Falsification, Revolution, and Continuity in the Development of Science', in: Suppes *et al.* (eds.), *Logic, Methodology, and Philosophy of Science* (Proc. of the IVth International Congress for Logic, Methodology, and Philosophy of Science in Bucharest, 1971), Warszawa-Amsterdam.

Krymskij, S.B.: 1965, 'Logical Principles of Transition from One Theory to Another', in: Kopnin and Popovitch (eds.), *The Logic of Scientific Investigation* (in Russian), Moscow.

Kuhn, T.S.: 1962, *The Structure of Scientific Revolution*, Chicago.

Kuhn, T.S.: 1969, Postscript to the 2nd edition of Kuhn, T.S.: 1962, *The Structure of Scientific Revolution*, Chicago.

Kuhn, T.S.: 1970, 'Logic of Discovery or Psychology of Research?', in: Lakatos and Musgrave (eds.), *Criticism and Growth of Knowledge*, Cambridge University Press.

Kuhn, T.S.: 1971, 'Notes on Lakatos', *BSPS*, Vol. VIII.

Kuptsov, V.I.: 1973, *Philosophy of Science* (in Russian), Moscow University Press.

Kuznietsov, B.G.: 1964, *Galileo* (in Russian), Moscow.

Kuznietsov, I.V.: 1948, *The Correspondence Principle in the Contemporary Physics and Its Philosophical Meaning* (in Russian), Moscow.

Kuznietsov, I.V.: 1968, 'Continuity, Unity, and Minimalization of Knowledge – Basic Features of the Scientific Method', in: *Materialist Dialectics and the Methods of the Natural Sciences* (in Russian), Moscow.

Kuznietsov, I.V.: 1970a, 'Engels' Conception of the Forms of Movement of Matter and Contemporary Natural Science', *VF*, No. 11.

Kuznietsov, I.V.: 1970b, 'Correspondence Principle', in: *Philosophical Encyclopedia* (in Russian), Moscow.

Laitko, H.: 1969, 'Das Korrespondenzprinzip als Methode der Theoretischen Erkenntnis', in: Laitko and Bellmann (eds.), *Wege des Erkennens*, Berlin.

Lakatos, I.: 1970, 'Falsification and the Methodology of Scientific Research Programmes', in: Lakatos and Musgrave (eds.), *Criticism and Growth of Knowledge*, Cambridge University Press.

Lakatos, I.: 1971, 'History of Science and Its Rational Reconstruction', *BSPS*, Vol. VIII.

Lakatos, I. (ed.): 1968, *The Problem of Inductive Logic*, Amsterdam.

Lakatos, I. and Musgrave, A. (eds.): 1968, *Problems in the Philosophy of Science*, London.

Lakatos, I. and Musgrave, A. (eds.): 1970, *Criticism and Growth of Knowledge*, Cambridge University Press.

Lassile, K.E.: 1971, 'Quantum Mechanics', in: *McGraw Hill Encyclopedia of Science and Technology*, New York.

Lenin, V.I.: 1909, *Materialism and Empirico-Criticism* (in Russian), Moscow.

Lenin, V.I.: 1933 (posthumously), *Philosophical Notebooks* (in Russian), Moscow.
Lenzen, V.: 1954, *Causality in Natural Science*, Berkeley.
Lewenstam, A.: 1974, 'Bohr's Correspondence Principle', in: Krajewski *et al.* (eds.), *Correspondence Principle in Physics and the Development of Science* (in Polish), Warszawa.
Lilley, S.: 1953, 'Cause and Effect in the History of Science', *Centaurus*, Vol. 3, No. 1-2.
McLaughlin, A.: 1971, 'Method and Factual Agreement in Science', *BSPS*, Vol. VIII.
McMullin, E.: 1970, 'The History and Philosophy of Science: A Taxonomy', *MSPS*, Vol. 5.
Majewski, Z.: 1967, 'The Place of Physics Among the Natural Sciences', *SF*, No. 3 (50).
Majewski, Z.: 1974, *The Dialectic of the Structure of Matter* (in Polish), Warszawa.
Mamchur, E.D.: 1971, 'The Criteria for the Theoretical Conceptions to Be Scientific', *VF*, No. 7.
Mamchur, E.D.: 1973, 'Value Factors in the Cognitive Activity of a Scientist', *VF*, No. 9.
Mamchur, E.D. and Illarionov, S.V.: 1973, 'The Regulative Principles of the Theory Construction', in: Omielianovskij *et al.* (eds.), *Synthesis of the Contemporary Scientific Knowledge* (in Russian), Moscow.
Manasian, A.S.: 1973, *The Problem of Development of the Scientific Knowledge* (in Russian), Erevan.
Mannheim, K.: 1929, *Ideologie und Utopie*, Bonn.
Marx, K.: 1859, *Zur Kritik der politischen Ökonomie*, Berlin.
Marx, K.: 1867, *Das Kapital*, Band I, Hamburg.
Masterman, M.: 1970, 'The Nature of Paradigm', in: Lakatos and Musgrave (eds.), *Criticism and Growth of Knowledge*, Cambridge University Press.
Mayzel, B.M.: 1975, 'The Problem of Knowledge in Popper's Philosophical Papers in the 60's', *VF*, No.6.
Mejbaum, W.: 1964, 'Gesetze und Ihre Formulierungen' (Polish original 1964), in: G. Kröber (ed.), 1968, *Das Gesetzesbegriff in der Philosophie und den Einzelwissenschaften*, Berlin.
Metzger, H.: 1926, *Les concepts scientifiques*, Paris.
Meyer-Abich, K.M.: 1965, *Korrespondenz, Individualität und Komplementarität*, Wiesbaden.
Meyerson, E.: 1908, *Identité et réalité*, Paris.
Meyerson, E.: 1931, *Du cheminement de la pensée*, Paris.
Mikulinskij, S.R.: 1974, 'Methodological Problems of the History of Science', in: Kedrov and Ovchinnikov (eds.), *Problems of the History and the Methodology of Scientific Knowledge*, Moscow.
Mikulinskij, S.R. and Markova, L.A.: 1971, 'On Different Conceptions of Factors of the Development of Science', *VF*, No. 8.
Miller, D.: 1975, 'The Accuracy of Predictions', *Synthese*, Vol. 30, No.1-2.
Miś, A.: 1974, 'On the Book "Elements of the Marxist Methodology of the Humanities"', *CiS*, No. 8 (109).
Misiek, J.: 1969, 'On the Metrics of Physical Space and Time', *SF*, No. 5 (60).
Nadel-Turoński, T.: 1974, 'Semantical Complementarity and a Conception of Correspondence', in: Krajewski *et al.* (eds.), *Correspondence Principle in Physics and the Development of Science* (in Polish), Warszawa.

Nagel, E.: 1961, *The Structure of Science*, New York.
Nagel, E.: 1970, 'Issue in the Logic of Reductive Explanation', in: Kieffer and Munitz (eds.), *Contemporary Philosophical Thought*, Vol. 2, 'Mind, Science, History', Albany.
Natorp, P.: 1910, *Die logischen Grundlagen der exakten Wissenschaften*, Leipzig.
Needham, J.: 1953, 'Thoughts on the Social Relations of Science and Technology in China', *Centaurus*, Vol. 3, No. 1-2.
Niedźwiedzki, W.: 1974, 'Theory, Correspondence, Correspondence Principle', in: Krajewski *et al*. (eds.), *Correspondence Principle in Physics and the Development of Science* (in Polish), Warszawa.
Nikitin, E.P.: 1972, *Explanation – a Function of Science* (in Russian), Moscow.
Novaković, S.: 1974, 'Is the Transition from an Old Theory to a New One of a Sudden and Unexpected Character?', *BSPS*, Vol. XIV.
Nowak, I.: 1974, 'Reduction and Correspondence', in: Krajewski *et al*. (eds.) *Correspondence Principle in Physics and the Development of Science* (in Polish), Warszawa.
Nowak, I.: 1975a, *Dialectical Correspondence in the Development of Science* (in Polish), Warszawa-Poznań.
Nowak, I.: 1975b, 'Idealization and the Problem of Correspondence', *PSPSH*, Vol. I, No. 1.
Nowak, L.: 1971a, *Foundations of Marxian Methodology* (in Polish), Warszawa.
Nowak, L.: 1971b, 'Galileo of Social Sciences' (in Polish), *Nurt*, No. 1 (69).
Nowak, L.: 1974a, *Principles of the Marxist Philosophy of Science* (in Polish), Warszawa.
Nowak, L.: 1974b, 'Relative Truth – Correspondence – Absolute Truth', in: Krajewski *et al*. (eds.), *Correspondence Principle in Physics and the Development of Science* (in Polish), Warszawa.
Nowak, L.: 1975, 'Idealization: a Reconstruction of Marx' Ideas', *PSPSH*, Vol. I, No. 1.
Nowak, S.: 1970, *The Methodology of the Social Sciences* (in Polish), Warszawa.
Nyssanbayev, A.: 1965, 'Correspondence Principle and Mathematics', *VF*, No. 7.
Nyssanbayev, A.: 1974, 'Correspondence Principle in Mathematics', in: Kedrov and Ovchinnikov (eds.), *Problems of the History and the Methodology of Scientific Knowledge* (in Russian), Moscow.
Omielianovskij, M.E., Fock, V.A., and Barashenkov, V.S. (eds.): 1973, *Synthesis of the Contemporary Scientific Knowledge* (in Russian), Moscow.
Oppenheim, P. and Putnam, H.: 1958, 'Unity of Science as a Working Hypothesis', *MSPS*, Vo. II.
Ovchinnikov, N.F.: 1974, 'The Features of Development and Tendencies towards Unity of the Scientific Knowledge', in: Kedrov and Ovchinnikov (eds.), *Problems of the History and the Methodology of Scientific Knowledge* (in Russian), Moscow.
Patrays, W.: 1976, *Experiment and Idealization* (in Polish), Warszawa-Poznań.
Pearce, G. and Maynard, P. (eds.): 1973, *Conceptual Change*, Dordrecht.
Petersen, A.: 1968, 'On the Philosophical Significance of the Correspondence Argument', *BSPS*, Vol. V.
Pietruska-Madej, E.: 1974, 'The Postulate of Correspondence and Its Functioning in the History', in: Krajewski *et al*. (eds.) *Correspondence Principle in Physics and the Development of Science* (in Polish), Warszawa.
Pietruska-Madej, E.: 1975, *Methodological Problems of the Chemical Revolution* (in Polish), Warszawa.

Pietruska-Madej, E.: 1977, 'Anti-Cumulative Changes in Science', in: Krajewski *et al.* (eds.), *Relations Between Theories in the Course of the Growth of Science* (in Polish), Wrocław.

Poincaré, H.: 1902, *La science et la hypothèse*, Paris.

Poincaré, H.: 1906, *La valeur de la science*, Paris.

Polikarov, A.: 1966, *Relativity and Quanta* (in Russian), Moscow.

Polikarov, A.: 1973, *Science and Philosophy*, Sofia.

Popovitch, M.V.: 1969, 'On the Universality of the Logic', *VF*, No. 7.

Popper, K.R.: 1935, *Logik der Forschung*, Wien.

Popper, K.R.: 1957, 'The Aim of Science', in: Popper, 1972, *Objective Knowledge* Oxford.

Popper, K.R.: 1963, *Conjectures and Refutations,* London.

Popper, K.R.: 1970, 'Normal Science and Its Dangers', in Lakatos and Musgrave (eds.), *Criticism and Growth of Knowledge*, Cambridge University Press.

Popper, K.R.: 1972, *Objective Knowledge*, Oxford.

Post, H.R.: 1971, 'Correspondence, Invariance and Heuristics: in Praise of Conservative Induction', *Stud. Hist. Phil. Sci.*, Vol. 2, No.3.

Putnam, H.: 1965, 'How Not to Talk About Meaning', *BSPS*, Vol. II.

Putnam, H.: 1973, 'Explanation and Reference', in: Pearce and Maynard (eds.), *Conceptual Change*, Dordrecht.

Rainko, S.: 1967, 'Diachronic Epistemology', *SF*, No. 1 (48).

Rainko, S.: 1971, *The Role of Subject in Knowledge* (in Polish), Warszawa.

Rainko, S.: 1972, 'K. Mannheim's Conception of Epistemology', *SF*, No. 7-8 (80-81).

Reichenbach, H.: 1946, *Philosophical Foundations of Quantum Mechanics,* University of California Press.

Reichenbach, H.: 1951, *The Rise of Scientific Philosophy*, University of California Press.

Rodnyj, N.I.: 1966, 'On the Role of Contradictions in the Development of Natural Science' (in Russian), *Organon*, No. 3.

Rodnyj, N.I.: 1973, 'The Problem of Scientific Revolution in the Conception of T. Kuhn', in: *Conceptions of Science in the Bourgeois Philosophy and Sociology* (in Russian), Moscow.

Rozental, M.M.: 1968, *Dialectic of 'The Capital'* (in Russian), Moscow.

Rubinowicz, A.: 1961, 'Correspondence Principle', in: *Encyclopedic Dictionary of Physics*, Vol. 2, Oxford.

Rudner, R.S.: 1966, *Philosophy of Social Sciences*, Engl.Cliffs, N.J.

Ruppert-Hall, A.: 1962, *The Scientific Revolution 1500-1800*, London.

Schaff, A.: 1951, *Problems of the Marxist Theory of Truth* (in Polish), Warszawa.

Schaffner, K.F.: 1967, 'Approaches to Reduction', *PS*, Vol. 34, No. 2.

Scheibe, E.: 1973, 'The Approximative Explanation and the Development of Physics', in: Suppes *et al.* (eds.), *Logic, Methodology, and Philosophy of Science*, Warszawa-Amsterdam.

Sellars, W.: 1973, 'Conceptual Change', in: Pearce and Maynard (eds.), *Conceptual Change*, Dordrecht.

Shakhparonov, M.I.: 1951, 'Review of I.V. Kuznietsov 1948' in *Viestnik Moskovskovo Universiteta*, No. 3.

Shea, W.R.: 1972, *Galileo's Intellectual Revolution*, London.

Shvyriov, V.S.: 1971, 'Analysis of Scientific Knowledge in the Contemporary Philosophy of Science', *VF*, No. 2.

Siemek, M.: 1974, 'Marxism and Hermeneutic Tradition', *SF*, No. 11 (108).

Siemens, W.D.: 1971, 'A Logical Empiricist Theory of Scientific Change?', *BSPS*, Vol. VIII.

Ślęczka, K.: 1972, 'Praise of the Idealization (Review of L. Nowak 1971)', *SF*, No. 9(82).

Solla Price, de D.J.: 1961, *Science since Babylon*, New Haven.

Subbotin, A.L.: 1964, 'Idealization as a Method of Scientific Knowledge', in: P. Tavaniets (ed.), *Problems of Logic of Scientific Knowledge* (in Russian), Moscow.

Stakhanov, I.P.: 1974, 'The Evolution of Physical Theories', in: Kedrov and Ovchinnikov (eds.), *Problems of the History and the Methodology of Scientific Knowledge* (in Russian), Moscow.

Stanosz, B.: 1972, 'Review of Lakatos and Musgrave (eds.) 1970', *ZN*, Vol. VIII, No. 2.

Strauss, M.: 1970, 'Inter-Theory Relations', in: Weingartner and Zecha (eds.), *Induction, Physics and Ethics*, Dordrecht.

Such, J.: 1969, 'Jevons about Laws of Nature', in: W. Krajewski (ed.), *Concept of Law of Science at the End of the XIXth Century* (in Polish), Wroclaw.

Such, J.: 1972a, *On the Universality of Laws of Science* (in Polish), Warszawa.

Such, J.: 1972b, 'The Marxian Method of Abstraction and Gradual Concretization in Natural Science', *SF*, No. 2 (75).

Such, J.: 1974, 'Correspondence Relation and Entailment', in: Krajewski *et al.* (eds.), *Correspondence Principle in Physics and the Development of Science* (in Polish), Warszawa.

Such, J.: 1975, *Problems of the Verification of Knowledge* (in Polish), Warszawa.

Suppes, P., Henkin, L., Yoja, A., and Moisil, G.C. (eds.): 1973, *Logic, Methodology, and Philosophy of Science* (Proceedings of the IVth International Congress for Logic, Methodology, and Philosophy of Science in Bucharest, 1971), Warszawa-Amsterdam.

Suszko, R.: 1957, 'Formal Logic and Some Problems of Epistemology', *MF*, No. 2-3 (28-29).

Suszko, R.: 1968, 'Formal Logic and the Development of Knowledge', in: Lakatos and Musgrave (eds.), *Problems in the Philosophy of Science*, London.

Synowiecki, A.: 1969, *Problem of Mechanistic Philosophy in Natural Sciences* (in Polish), Wroclaw.

Szmatka, J.: 1975, 'Problem of Theoretical Reduction in the Philosophy of Science and Sociology', *SF*, No. 3 (112).

Szumilewicz, I.: 1969, 'L. Boltzmann's Mechanism and the Postulate of Microreduction', in: *Philosophical Essays in Honor of T. Czezowski* (in Polish), Toruń.

Szumilewicz, I.: 1970, 'The Postulate of Microreduction since Lucretius till Boltzmann', *KHNT*, Vol. XIV, No. 1.

Szumilewicz, I.: 1974, 'Correspondence Principle and the Problem of Incommensurability of Theories', in: Krajewski *et al.* (eds.) *Correspondence Principle in Physics and the Development of Science* (in Polish), Warszawa.

Toulmin, S.: 1961, *Foresight and Understanding*, London.

Toulmin, S.: 1970, 'Does the Distinction between Normal and Revolutionary Science Hold Water?', in: Lakatos and Musgrave (eds.), *Criticism and Growth of Knowledge*, Cambridge University Press.

Tuomela, R.: 1973a, 'Deductive Explanation of Scientific Laws, in: Bunge (ed.), *Problems, Tools, and Goals*, Dordrecht.

Tuomela, R.: 1973b, 'Theoretical Concepts', *LEP*, No. 10.

Twardowski, K.: 1900, *Über sogenannte relative Wahrheiten*, Archiv für systematische Philosophie VIII, Berlin 1902 (Polish original 1900).

Wartofsky, M.W.: 1968, *Conceptual Foundations of Scientific Thought*, New York.

Watkins, J.: 1970, 'Against "Normal Science"', in: Lakatos and Musgrave (eds.), *Criticism and Growth of Knowledge*, Cambridge University Press.

Weber, M.: 1922, *Gesammelte Aufsätze zur Wissenschaftslehre*, Tübingen.

Weingartner, P. and Zecha, G. (eds.): 1970, *Introduction, Physics, and Ethics*, Dordrecht.

Williams, L.P.: 1968, 'Epistemology and Experiment: The Case of Michael Faraday', in: Lakatos (ed.), *The Problem of Inductive Logic*, Amsterdam.

Williams, L.P.: 1950, 'Normal Science, Scientific Revolutions and the History of Science', in: Lakatos and Musgrave (eds.), *Criticism and Growth of Knowledge*, Cambridge University Press.

Witt-Hansen, J.: 1965, 'Two Methods of Justification and the Correspondence Principle', in: Ajdukiewicz (ed.), *The Foundation of Statements and Decisions*, Warszawa.

Wòjcicki, R.: 1974, *Formal Methodology of the Empirical Sciences* (in Polish), Wroclaw.

Yaroshevskij, M.G.: 1974, 'The Structure of Scientific Activity', *VF*, No. 11.

Zamiara, K.: 1974, *The Methodological Importance of the Strife on the Cognitive Status of Theories* (in Polish), Warszawa.

Zhdanov, I.A.: 1968, 'Development of Matter and the Organic Chemistry', in: *Problems of Development in the Contemporary Natural Sciences* (in Russian), Moscow.

Zilsel, E.: 'The Development of Rationalism and Empiricism', *International Encyclopedia of United Science*, Vol. II, No. 8, Chicago.

Żytkow, J.: 1974, 'The Structure of Physical Theory and the Relations of Reduction and Correspondence', in: Krajewski *et al.* (eds.), *Correspondence Principle in Physics and the Development of Science* (in Polish), Warszawa.

INDEX OF NAMES